STORY GENIUS

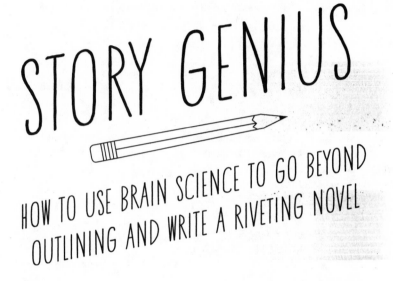

STORY GENIUS

HOW TO USE BRAIN SCIENCE TO GO BEYOND OUTLINING AND WRITE A RIVETING NOVEL

[*Before You Waste Three Years Writing 327 Pages That Go Nowhere]

LISA CRON

TEN SPEED PRESS
Berkeley

Published in the United States by Ten Speed
Press, an imprint of the Crown Publishing
Group, a division of Penguin Random House
LLC, New York.
www.crownpublishing.com
www.tenspeed.com

Ten Speed Press and the Ten Speed Press
colophon are registered trademarks of Penguin
Random House LLC.

Library of Congress Cataloging-in-Publication
Data is on file with the publisher.

Trade Paperback ISBN: 978-1-60774-889-2
eBook ISBN: 978-1-60774-890-8

Printed in the United States of America

Design by Debbie Berne

20 19 18 17 16 15 14 13 12 11

First Edition

For Daisy,
whose story is just beginning.

CONTENTS

INTRODUCTION

What's the biggest mistake writers make? This is the question I've been asked most frequently over the years. The answer is easy: they don't know what a story is. So even though they have a great idea, their prose is gorgeous, and there's a lot of action, there's no real story, and so no driving sense of urgency, which translates to: no readers.

The result? Countless writers end up heartbroken because no matter how hard they work, how many writing workshops they take, how many degrees they earn, they still can't get an agent, can't get a book deal, and if they decide to self-publish (in order to show those talent-blind publishing bigwigs a thing or two), they can't get anyone other than their friends and family to buy their book. The statistics can be scary. In 2012, the *New York Times* reported that most self-published books sell fewer than 150 copies; surveys reveal that agents reject over 96 percent of the submissions they receive (personally, I'd put the number even higher). So it's no surprise that writers end up disappointed, sad, and sometimes even a tad bitter. Worse, they're convinced that their failure proves one thing only: they have no talent.

That's when that internal voice we all have, the one that pretends to have our best interest at heart, moves in for the kill. *Whatever made me think I could be a writer? I should give it up immediately and express my creativity some other way. Like—I don't know—interpretive dance.* Don't! Not only because chances are you *do* have the talent, but seriously, the world has way too many interpretive dancers as it is.

Here's the truth: not understanding how story really works is not your fault. It's on a par with not knowing exactly how your body absorbs the nutrients in the food you eat. You know that it does, and if you took high school biology, you probably remember that it has something to do with cells and membranes and amino acids, but the how of it is invisible (thank heaven). The same is true of the effect that story—all stories—have on you, and even more surprising, *why* they have that effect.

This book is designed to help you crack the story code, and make what was invisible, visible—not to mention eminently doable. It will turn you into a Story Genius. It will show you, step by step, how to craft a blueprint for your story that will set you up for success from the get-go. It will also drastically cut down on rewriting time—and it's the only thing that will. You'll not only hook readers from the very first sentence, but your novel—or screenplay, play, or short story—will be deeper, richer, and more compelling than anything you've written before. How can I be sure? Because we're not talking about some new flavor-of-the-month writing system conjured out of thin air. We're talking brain science.

Humans are wired for story. We hunt for and respond to certain specific things in every story we hear, watch, or read—and they're the exact same specific things, regardless of the genre. Why is this so? Because story is the language of the brain. We think in story. The brain evolved to use story as its go-to "decoder ring" for reality, and so we're really expert at probing stories for specific meaning and specific info—and I mean all of us, beginning at birth. Even a kindergartner recognizes an effective story, because it's built into the architecture of the brain. Story is how we make sense of the world around us; it's a system that predates written language by eons. Heck, before spoken language, we grunted and signed in story. I'd wager that early in the morning, the cranky among us still do.

Because our response to story is hardwired, it's not something we have to learn or even think about, which is why we are often unaware of the power story has over us. When a story grabs you, you're in its sway, no questions asked. You may have heard the oft-expressed sentiment that getting lost in a good story demands a "willing suspension of disbelief." In fact, this couldn't be less true, because it implies we have a choice as to whether we fall under the spell of a captivating story. We don't have a choice. The power story has over us is biological. But while *responding* to story is hardwired, creating a story is not. As the great Southern writer

Flannery O'Connor once noted, "Most people know what a story is until they sit down to write one."[1] But here's the part she missed: before we can learn to write a story, we have to know what a story actually is. That is, we have to know what's really hooking and holding readers.

The problem is that most writers mistake story for the things we can see on the page: the stunning prose, the authoritative voice, the intense and exciting plot, the clever structure. It's a very natural mistake, and a crippling one. Because while no one could deny that all those things are important, they lack the crucial element that gives a story meaning and brings it to life.

What drives a story forward is, at first blush, invisible. It's not talent. It's not voice. It's not the plot. Think electricity. The same way even the most powerful lamp is useless unless it's plugged in, a story can't engage readers without the electricity that illuminates the plot, the voice, and the talent, bringing them to life.

The question is: what, specifically, generates that juice?

The answer is: it flows directly from how the protagonist is making sense of what's happening, how she struggles with, evaluates, and weighs what matters most *to her*, and then makes hard decisions, moving the action forward. This is not a general struggle, but one based on the protagonist's impossible goal: to achieve her desire *and* remain true to the fear that's keeping her from it. As we'll explore in detail, story is not about the plot, or what happens. Story is about how the things that happen in the plot *affect* the protagonist, and how he or she changes internally as a result.

Think of the protagonist's internal struggle as the novel's live wire. It's exactly like the third rail on a subway train—the electrified rail that supplies the juice that drives the cars forward. Without it, that train, no matter how well constructed, just sits there, idling in neutral, annoying everyone, especially at rush hour. Ultimately, all stories are character driven—yes, *all* stories, including *50 Shades of Grey, A Is for Alibi, Die Hard, War and Peace, The Goldfinch,* and *The Little Engine That Could.*

In a novel, everything—action, plot, even the "sensory details"— must touch the story's third rail in order to have meaning and emotional impact. Anything that doesn't impact the protagonist's internal struggle, regardless of how beautifully written or "objectively" dramatic it is, will stop the story cold, breaking the spell that captivated readers, and unceremoniously catapulting them back into their own lives.

The reason that the vast majority of manuscripts are rejected—either by publishers or by readers—is because they do not have a third rail. This is where writers inadvertently fail. This is the biggest mistake they make. And so they write and rewrite and polish an impressive stack of pages in which a bunch of things happen, but none of it really matters because that's all it is—a bunch of external things that the reader has no particular reason to care about.

Story is about an internal struggle, not an external one. It's about what the protagonist has to learn, to overcome, to deal with *internally* in order to solve the problem that the *external* plot poses. That means that the internal problem predates the events in the plot, often by decades. So if you don't know, specifically, what your protagonist wants, what internal misbelief is standing in his way—and most important, why—how on earth can you construct a plot that will force him to deal with it? The answer is simple: you can't.

This is why you have to know everything there is to know about the protagonist's specific internal problem *before* you create the plot, and why this knowledge will then, with astonishing speed, begin to generate the plot itself. Story first, plot second, so that your novel has the juice to instantly captivate your readers, biologically hooking them before they know what hit 'em.

That's the power *Story Genius* will give you. It will take you, step by step, from the first glimmer of an idea, to an evolving, multilayered cause-and-effect blueprint that transforms into a first draft with the authority, richness, and command of a fully realized sixth or seventh draft.

You'll notice I use the word "blueprint" throughout this book rather than "outline." That's because in writing parlance the term *outline* typically refers to a scene-by-scene summary of the external plot—the surface of the novel. That is not what this book is about. We're going beneath the surface to where the real story lies—the story that the reader's brain is wired to find irresistible. The blueprint we're talking about in *Story Genius* is not a general outline of the things that happen in the plot; it's a fully realized synthesis of the internal and external layers of your story from beginning to end. You will begin to write your novel as you blueprint—in fact, much of what is in your blueprint will be in your novel. Nothing in this process goes to waste. None of it is "prewriting." The result? A riveting novel that will change how your readers see the world.

PART I

WHAT A STORY IS, AND WHAT IT ISN'T

1

STORY: THE BRAIN'S DECODER RING

*There have been great societies that did
not use the wheel, but there have been no
societies that did not tell stories.*

—URSULA K. LEGUIN

Pop quiz: It's been a long day and you're looking for a way to kick back and relax. Which of the following choices are biologically guaranteed to mute all those nagging real-world worries and make you feel pretty darn great by temporarily changing your body chemistry?

1. A nice glass of Pinot Noir
2. A box of chocolates
3. A novel

The answer is all three. But a novel is by far the most potent drug, the longest lasting, and the only one that won't leave you with any regrets in the morning. Well, except maybe one.

Imagine this: You're finally ready to tumble into bed, glad to be turning in early because you have a big meeting in the morning and have to be up at the crack of dawn. You reach for the novel on your nightstand. You figure you'll read a chapter—you know, to relax—and then, lights out. But when you get to the end of the chapter, you're thinking, *"Wait, what will Priscilla do when she finds the note Kendrick left for Bridgette? She's sure*

9

to misread it and . . . ." So you decide to read *one* more page, just to find out. And one page turns into three, which turns into ten. Suddenly you're not tired. In fact the entire concept of "tired" has ceased to make sense. The real world has vanished, and you're in a nice comfy bubble floating somewhere in space. It's as if someone pressed the pause button on your own life, allowing you to live and breathe in an alternate reality. Priscilla's reality. The pages keep flying by, until you notice that there's an annoyingly bright light coming in beneath the blinds. Has someone parked a Mack truck outside your window? Then it hits you in a wave of panic: it's dawn. You've stayed up reading all night. It's about then that you remember with stunning clarity exactly what tired means.

It happens to all of us. But why? You knew you had to get up early, and that a lack of sleep tends to leave you cognitively depleted and, if you're anything like me, kind of grouchy. So why on earth did you keep on reading? Before you beat yourself up for not having the internal fortitude to stick to the plan and put the damn book down after a page or two, consider this: an effective story is, literally, an offer your brain can't refuse. You didn't *decide* to keep reading—it was a biological reaction. Nature made you do it. Of course, I wouldn't recommend that line of explanation to your boss, should she catch you dozing off during the meeting or, worse, bursting into tears when someone mentions that your socks don't match. Since we don't understand the power story has over us, let's face it, she'd think you were nuts.

That's the scary thing about story. We're bewitched and affected by stories every minute of every day whether we know it or not. But like your well-meaning boss, most of us are completely unaware of the hardwired power story has on us. For writers, though, this is where the keys to the kingdom lie. Understanding why stories are so influential, and exactly what it is that gives them the ability to transfix and then transform readers' lives, will allow you to wield that power in your novel.

That's why before you can develop an effective blueprint (or if you've already started your novel, before you write another word), you need to know what your brain is *really* responding to when it filters out the real world in order to dive headfirst into the world on the page. To that end, in this chapter we'll look at the hardwired purpose of story; examine how story and the brain evolved in tandem; discuss what gives story its unparalleled power over us; and explore what a story actually *is*, based on what the brain is wired to crave, hunt for, and respond to in every story we hear.

The Hardwired Purpose of Story

Why *do* normally responsible adults like you and me check out of reality so completely when we're under the spell of a compelling story? That's something evolutionary biologists have been wondering about for a long time, and with good reason, because staying up all night to finish reading that novel was definitely counterproductive. But hey, at least you survived to see the dawn. Back in the Stone Age, making it through the night was a much dicier proposition, and putting reality on hold for even a moment left you vulnerable to all sorts of pouncing predators, human or otherwise. In other words, getting lost in a story could be deadly, which is why scientists figured there had to be a damn good reason for it, or else natural selection would have weeded out those of us prone to getting lost in a story faster than you can say, "Just one more chapter, I promise!"

There is a damn good reason. Story was the world's first virtual reality. It allowed us to step out of the present and envision the future, so we could plan for the thing that has always scared us more than anything: the unknown, the unexpected. What better way to figure out how to outsmart those potential pouncing predators before they sneak up behind you?

Sure, being in the "now" is a good idea *sometimes*, but if you were *always* in the now you wouldn't even know there was a tomorrow, let alone be able to speculate on the dangers and delights that might be lurking there. Stories let us vicariously try out difficult situations we haven't yet experienced to see what it would *really* feel like, and what we'd need to learn in order to survive. So it's no surprise that there's never been a society on earth that didn't have storytelling. It's a human universal, which probably should have clued us into the fact that there might be a wee bit more to it than just a great way to spend a rainy Saturday afternoon or a long night before a big meeting.

How Story Hacks the Reader's Brain

But if story has so much power, if it's so critical to our well-being, why do we tend to brush it off as mere entertainment? Why do we think that losing ourselves in a good story is something that's optional—a treat we give ourselves at the end of a long day of actual work, when we want to leave the trials and tribulations of the real world behind and plunge into the refreshing world of "make believe"? Why, indeed, does no less an authority than the *Oxford English Dictionary* define story as "An account of imaginary or real people and events told for entertainment"?

The answer is simple. We've mistaken the feeling story gives us—that deliciously seductive pleasure—for its purpose. And as with any seduction, once we're under its spell, only one thing matters: this fabulous, enthralling moment, *right now.* Did you ever consider the consequences of falling sway to a particular story's magic? Hell, the fact that there might even *be* a consequence is completely off our radar. We all know that once the story ends, it's over, and we're back in real life, the exact same person we were before we started reading. I mean, it's not like stories affect us in any kind of quantifiable way, right? Wrong!

Stories feel good for the same reason food tastes good and sex feels good: because without them we couldn't survive. Food nourishes us, sex begets us, stories educate us. It's just that with food and sex, the consequences tend to become apparent pretty darn quickly, so we're well aware of their possible after-effects from the get-go—knowledge that *still,* I might point out, tends to go missing in the face of a nice big ice cream sundae or a longed-for come-hither glance.

Since we don't see the same kind of clear-cut potential consequence that reading a story has on our lives, it's easy to imagine there isn't one. But story's effect on us is just as profound, as life-altering, and as biologically driven. It turns out that great feeling you get when you're lost in a good story, the feeling that can keep you up all night reading, is not ephemeral, it's not arbitrary, it's not pleasure for pleasure's sake, it's not even *the point*. It's actually the biological lure, the hook that paralyzes you, making the real world vanish so you can experience the world of the story. That feeling is what compels us to drop everything and pay attention.

What actually causes that great feeling is a surge of the neurotransmitter dopamine. It's a chemical reaction triggered by the intense curiosity that an effective story always instantly generates. It's your brain's way of rewarding you for following your curiosity to find out how the story ends, because you just might learn something that you need to know.

When we're under the spell of a compelling story, we undergo internal changes along with the protagonist, and her insights become part of the way we, too, see the world. Stories instill meaning directly into our belief system the same way experience does—not by telling us what is right, but by allowing us to *feel* it ourselves. Because just like life, story is emotion based. As Harvard psychology professor Daniel Gilbert said, "Indeed, feelings don't just matter, they are what mattering means."[1] In life, if we can't feel emotion, we can't make a single rational decision—it's biology. In a story, if we're not feeling, we're not reading. It is emotion, rather than logic, that telegraphs meaning, thus emotion is what your novel must be wired to transmit, straight from the protagonist to us.

We're wired to crave, hunt for, and latch onto what the protagonist feels, so that we can experience his struggles as if they were our own. When we're lost in a novel, the protagonist's internal struggle becomes ours, as do his hard-won truths. This is not a metaphor, but a fact. According to Jonathan Gottschall, author of *The Storytelling Animal*, functional MRI (fMRI) studies reveal that when we're reading a story, our brain activity isn't that of an observer, but of a participant.

Don't believe it? Go watch a horror movie at your local theater (soon, or there might not be any theaters left to go to). Now, when the monster is slashing at some poor, helpless victim who's trying desperately to get away, turn around and look behind you. As Gottschall says, chances are you'll see "the audience squirming in their seats. They pull in their elbows and scrunch up their knees, balling up to protect their vital organs." It's hard to watch their reaction without laughing. Clearly these are smart people who've seen a movie before. Why on earth would they think that they have to protect themselves from what's happening on the screen? The answer is, they're not thinking at all. They're experiencing it as if it were happening to them. For as Gottschall goes on to point out, "Their brains are instructing their bodies to do all the things they'd do if they were actually under mortal attack."[2]

That's the hardwired power that story has over us. It's pretty impressive, isn't it? Our brain registers an effective story the same way it registers the things that actually do happen to us out here in the good old analog world. This is what allows us to instantly channel the info we're always on the hunt for. To wit: we come to every story we hear—not just novels, which, evolutionarily speaking, arrived on the scene about five seconds ago—hardwired to ask one question in what's known as our cognitive unconscious: *What am I going to learn here that will help me not only survive, but prosper?*

We're not just talking about survival in the physical world but in the social world as well. Because contrary to popular belief, our need to belong to a group is just as biologically driven as is our need for food, air, and water. Once we'd learned to navigate the physical world, nature realized that if we were going to thrive, we needed to learn to work together. And so the ability to do just that was woven into our neural wiring about two hundred thousand years ago, when our brain had its last big growth spurt. We've long assumed that our big brains evolved to enable abstract reasoning, but science is now discovering that growth wasn't simply to make room for analytic ability. It was to expand our social cognitive skills, thus enabling us to do that thing we've all been encouraged to do since kindergarten: work well with others.

After all, even the most brilliant among us—that select few who come up with ingenious inventions that change everything—can't do it all by themselves. Can you imagine Steve Jobs, alone in that garage of his, assembling every Apple product ever made? That is, having first manufactured the parts all by himself, gotten the word out by himself, and then, once he'd amassed a slew of orders, hopping into a truck and delivering them by himself, after, of course, he built the road. And the truck. The point is, even the sharpest knives in the drawer can't cut it alone. We're *all* people who need people. Which means that in order to have a shot at prospering, we need to understand other people.

And so story's purpose evolved from simply decoding the mysteries of the physical world to deciphering the far more intricate social world—a far trickier task. After all, we can *see* the physical world, and learning things like "no matter how friendly the lion looks, *don't* pet it" is pretty straightforward. But when it comes to survival in the social world, the terrain is largely invisible. Sure, we can see what people do, but knowing *why* they're doing it—which is what matters most—is elusive, for the obvious reason that the answer is tucked into the one place that not even the NSA (or Mark Zuckerberg's minions) have yet found a way to tap: what someone else is actually thinking. What *do* they believe and why do they believe it? That's what we're dying to know, and what we're wired to respond to in every story we hear, especially novels. Novelists are revered (and at times feared) because they have been able to go deeper than scientists, shedding light on the inner reaches of someone else's mind, giving us insight into what makes people tick.

Understanding the motivation behind what someone does is what gives it meaning. Otherwise, when we give our faraway beloved a call, and with a heavy sigh he says, "I was happier before you rang," we might panic, thinking he's lost interest, when what he actually means is, *"Hearing your voice made me miss you, and that makes me sad I'm not with you."* In other words, the "why" is often very, very different from what it appears to be on the surface, and without some way to intuit what's really going on, how do we know whether to reactivate our membership on Match.com or to rejoice? We want to know the same thing that Nick, the dodgy husband in Gillian Flynn's *Gone Girl,* pensively wonders of his rather enigmatic wife on the novel's first page: "What are you thinking, Amy?" But how do we figure that out, short of going to a psychic? Do we do it, as Nick then contemplates, by "unspooling her brain and sifting through it, trying to catch and pin down her thoughts"? God, I hope not. Luckily (for our beloved) we have an even better alternative: we can sharpen our mind-reading skills by devouring a ton of novels. The purpose of story—of every story—is to help us interpret, and anticipate, the actions of ourselves and of others. And you have to admit, it's a far less messy alternative than all that unspooling.

The takeaway is: We don't turn to story to escape reality. We turn to story to navigate reality.

And if there's one thing that the reader's brain is not hardwired to be curious about, it's this: *I wonder what scrumptious metaphor the writer is going to use next. What breathtaking turn of phrase will she come up with now?* It turns out the brain is far less picky about lyrical language than we've been led to believe. We don't come for beautiful language, poetic writing, or even dramatic plot points. We come for something much deeper, much more meaningful: inside info on how to survive in this glorious, cruel, beautiful world, and in style no less. It's this inside info that we're wired to respond to, and it's this that any writer who wants to capture a reader's attention must master. But to do that, the first question we need to answer is: okay, what *is* a story?

What a Story Really Is

First, ditch the *Oxford English*'s definition of story as something created for entertainment. It could not be more wrong. It's like saying the sole purpose of food is to be delicious. What's more, like most general definitions, it's also utterly vague. Their definition basically boils down to this: a story is whatever someone finds entertaining—which could be, well, anything. Not very helpful for writers, is it? But there is one thing those British scholars got right: stories must indeed entertain, just like food has to taste good, otherwise no matter how nutritious it is, it gets shoved to the back of the fridge, next to the container of moldering kale. If novels didn't entertain us, we wouldn't pay attention to them, and they'd molder unread on the shelf.

What, then, *is* entertaining about stories? What is it that hooks us when we read, if not the beautiful language or the dramatic events? If curiosity is the key, if that's what gets the dopamine flowing, what is it that we're curious about? What *is* a story?

In a nutshell: A story is about how the things that happen affect someone in pursuit of a difficult goal, and how that person changes internally as a result.

Now, let's crack that shell wide open.

What happens in the story is the plot, the surface events of the novel. It is not the same thing as what the story is about. Not by a long shot.

The *someone* is the protagonist, and as we'll see, everything that happens in the plot will get its meaning and emotional weight based on how it affects her—not in general, but in pursuit of a difficult goal.

The *difficult goal* is, at its most basic, what's known as the *story problem*. All stories revolve around how someone solves a single, escalating problem they can't avoid. After all, if it were easy, it wouldn't be a problem, and there wouldn't be a story. It's not merely a surface problem, either, but one that causes the protagonist to struggle with a specific internal conflict at every turn, so that at the end she sees things quite differently than she did at the beginning.

And that *internal change?* That, my friends, is what the story is actually about: how your protagonist's external dilemma—aka the plot—changes her worldview.

And just to be clear: if you're writing a thriller or a mystery—whether a cozy, a hard-boiled police procedural, a courtroom drama, or anything in between—you do not get a pass on the inner story layer. After all, the goal isn't simply to have your investigator find out who did it; it's to find out how and why. And *why* is always internal. Think: What's the real reason for the crime, the hard-won answer to *"What was he thinking?"* In fact, mysteries of all genres tend to have a more complex inner layer, because each key character—the perp, the victim, the investigator—is driven by his or her internal agenda, and viewed through his or her subjective lens.

Story is about what happens internally, not externally. Not fully grasping the importance of this is what tanks countless novels. We don't come to story simply to watch the events unfold; we come to experience them through the protagonist's eyes, as she struggles with what to do next. That is what mesmerizes us: it's what we're curious about, it's what gives us the inside info we're hungry for. Yep, the protagonist's internal struggle is the story's third rail, the live wire that sparks our interest and drives the story forward.

The maddening catch for writers is that although responding to a story is something we can do from birth without thinking, the ability to *write* a story capable of hijacking the reader's brain is not part of the package. Although many of us are ace social storytellers—think gossip, anecdotes, and a riveting rendition of what happened on that date from hell—when it comes to creating a whole fictionalized world from scratch? That's a different skill altogether, and not standard-issue equipment in the human brain.

Then again, if the ability to write a compelling story were just as hard-wired as the ability to recognize one, we'd all be famous novelists, sitting by the pool with a flock of eager assistants fielding foreign rights offers, movie deals, and lucrative speaking gigs. Except it would be kind of hard to hire a staff, because *they'd* all be famous novelists, too. Plus, who would have time to read all those great novels?

As I'm sure you've figured out already, writing a novel is kind of hard. That's a given. But it's not nearly as difficult as it often feels. What undoes so many writers right out of the starting gate is something that seems so totally reasonable that it never occurs to us to question it: we decide that the first thing to do is to learn to write well. The trouble is, in learning to write well, we completely miss the boat, storywise.

2

MYTHS GALORE: EVERYTHING WE WERE TAUGHT ABOUT WRITING IS WRONG

The difficulty of literature is not to write,
but to write what you mean.
—ROBERT LOUIS STEVENSON

To hook a reader, you need to learn to write well. That's what we're taught in great detail from kindergarten on. The tacit promise is this: become a wordsmith and you automatically become a storyteller. Over and over we're told that the mechanics, conventions, and techniques of writing are what good writing is all about. How could you not believe it? Beautiful sentences are constantly praised. Misplaced commas are criticized. No one talks about story—which, as we have just seen, is the very thing that readers are hardwired to respond to. *Story* is treated as something elusive, something that magically appears when you nail the mechanics and learn to "write well."

What does "writing well" really mean? Here's a roundup of the usual suspects: it means coming up with great characters, interesting situations, dramatic scenes, intense conflict, compelling dialogue, gorgeous metaphors, and beautiful sentences, and then sprinkling in a lot of sensory details because that's what they say brings a story to life. That done, all you then have to do is unleash your creativity—cue the fairy godmother or the muse or a really inspiring dark night of the soul—and voilà! A story appears.

Sounds like a surefire recipe for success. Except for the small fact that it doesn't work. The conventions of writing—voice, structure, drama, plot, all of it—are the handmaiden of story, not the other way around. It's the story that gives those beautiful words, those interesting characters and all that drama, their power. It's the story that instantly sparks the reader's curiosity, triggering that irresistible sense of urgency that compels us to read on. Once we're securely hooked, it's the story that has the power to change how we view the world, and therefore what we go out and do in the world. That's why when it comes to effective writing, story is the only nonnegotiable. Everything else is gravy. To be clear, I'm not saying that learning to write well isn't a good idea. It is. But of the two, only story is essential. And, as we'll see throughout this book, understanding what a story is, and using that knowledge to develop a blueprint, will teach you everything you need to know to hook and hold your readers.

Before we can even think about blueprinting, it's crucial to dismantle a plethora of writing myths that, for all their good intentions, will lead you astray. These myths are dangerous because they are deep-rooted, widespread, and alluring, but the more you know about why they don't work, the easier it will be to resist them. In this chapter, we'll discover why both "plotting" and "pantsing" lead you away from the story, rather than to it; why beautiful prose means nothing in and of itself; how "writing well" can actually alienate the reader, even in literary fiction; why and how using story structure models like *The Hero's Journey* undermines your story; and why, even when it comes to the scenes you will be writing as you blueprint, that "shitty first draft" matters much more than you think.

The Myth of Great Writing

Here's a sobering question that might shake up many a literary circle: If great writing is what the brain's hungry for, if it's the thing that captivates readers, then could the *Fifty Shades of Grey* trilogy really have sold—are you sitting down?—100,000,000 copies and counting? (Yes, that's a hundred million as of this writing, which means that by the time you're reading this, that number has risen as steadily as . . . well, you get the picture.)

Confronted with this fact, the skeptic has two choices: Simply dismiss those *hundred million people* as readers who just don't know any better, poor saps. Or wonder what the hell is going on. Maybe, just maybe, there's something more compelling out there than beautiful writing, something that trumps beautiful writing every time.

After all, what do readers say about *Fifty Shades of Grey?* "It's horribly written, but I can't put it down." No one, not even the book's diehard fans, say that the author, E. L. James, "writes well." It's true that *Fifty Shades* is horribly written—by "beautiful writing" standards, that is. There's no way to pretty up the fact that Anastasia Steele, the spunky heroine, says "holy crap" forty times. "Holy crap" is an expression you should say only once in your whole entire life. If you already said it back when you were four, well, you've used your quota. Worse, she says "inner goddess" so many times that if you used it for a drinking game, you'd be in rehab before you were halfway through the book. Is this beautiful writing? Not a chance. And yet, the year Random House acquired the trilogy it catapulted them into the black. In fact, they gave every employee in the United States—from top editors to warehouse workers—a five thousand dollar holiday bonus. Clearly something is going on here, something that has absolutely nothing to do with the "quality" of the writing. That something is story. It's irresistible, which probably explains why so many otherwise literary readers have secretly hunkered down with *Fifty Shades*—that is, late at night once they were sure everyone else in the house really was asleep, the blinds were securely drawn, and no one in their reading group was anywhere within a ten-mile radius.

And yet when we hunker down to write, given what we've been taught to focus on as *writers*, the power of story is rarely on our radar. Instead, we put our faith in the power of the beautiful words themselves to lure readers, thus mistaking the wrapping paper for the present.

But if it's not about beautiful writing, why do we so wholeheartedly believe that it is? There are two main reasons.

Biology Has Us at Hello

The first reason takes us straight back to biology. The initial job of an effective story is to anesthetize the part of your brain that knows it is a story. That's why when you're caught up in a story it doesn't *feel* like a story at all; it feels like reality—especially since you're experiencing what's happening to the protagonist along the same neural pathways you would if it were actually happening to you. Those of you willing to admit to reading *Fifty Shades* know exactly what I'm talking about. You're not just reading about Anastasia Steele—you *are* her. Which is why when you're engrossed in a story, the last thing you can do, or want to do for that matter, is figure out how the writer has created the palpable sense of reality that you're lost in. You just want to enjoy it; you just want to be in it.

But even in that state, there are a couple of things you *can* see, because, well, they're visible. The language, for example. That's why it's incredibly easy to mistake the author's voice for what has you hooked. How many times have you said to a friend, "I love that writer. She's so clever, so wry, so insightful. I'd read anything she writes!" The belief that this is, indeed, what's grabbed you is furthered when you then read a review of the novel in which the reviewer has pulled out a lovely sentence or two as proof of just how compelling the book is. Reading it, you swoon, thinking, *I want to learn to write luscious sentences just like that!* rather than *I want to learn to write the kind of story that would give sentences like that their power!*

There is one other thing you can see even when a story has you under its spell: the plot; that is, the events that take place, the external things that happen, the surface of the novel. After all, the plot is concrete, it's clear, it's visual. So, as you close the book with a satisfied sigh, the lesson seems obvious—a writer's goal is to learn to write beautiful sentences and figure out a plot, and then, if you have the talent, a story will appear.

For a very, very select few writers, that *is* how it works. Some people, in fact, are able to write enthralling novels in record time from the disheartening (for the rest of us) age of twenty-four. So it makes complete sense that we'd turn to them, hungry to learn their methods, their techniques, their processes, the better to crib them and make them our own. Which brings us to the second reason why we've been led to believe that a writer's first job is to learn to write well.

The Best Writers Sometimes Give the Worst Advice

Writing tends to be taught by very accomplished writers, many of whom have a natural aptitude for story from the get-go, the way some athletes are born with a physical prowess that the rest of us couldn't come close to no matter how many boxes of Wheaties we scarf down, or how much cross-training we sweat through. But as counterintuitive as it may sound, that doesn't make these writers good teachers. In fact, often the opposite is true, because we're much better at teaching something that we've learned through experience than we are at teaching things we innately know. When we innately know how to do something, we assume it's part of the standard operating package we're all born with. It's so much a part of us that we never even think about the "how" of it—it just *is*. For instance, we all know how to walk, but when we stand up with the intention of going into the kitchen for a snack, we never have to think, "Okay, let's see, first I'll move my right leg, then, if memory serves, I'll . . . ." Instead our legs seem to know what to do all on their own. How convenient is that?

But imagine if someone came up to you and said, "I've never walked before; can you give me a quick tutorial on the basics?" You'd say, "Of course, I walk all the time. You just put one foot in front of the other, like this . . ." and then you'd gracefully glide across the room, turning around just in time to catch them rolling their eyes. "Yes, yes," they'd say, "I can see that, I'm not blind. But *how* are you doing it? How do you know which muscle to move first?" At which point you'd stand there blinking. Can you explain what you're doing when you walk? Could you tell someone what muscles to move, how, and in what combination in order to take even a single step, let alone keep walking all the way to the fridge? It's an incredibly complex and well-orchestrated series of unconscious decisions, reactions, and electrical impulses that I'm betting the vast majority of us have never given a moment's thought to. It's muscle memory; we do it automatically.

It's the same with natural storytellers. They've never had to deconstruct what they're doing, or pinpoint what it is that the reader is really responding to. These lucky pups have such a natural sense of story that often the novel merely unfolds as they write, delightfully surprising them at every turn. It's not talent, or "the muse." It's that their cognitive unconscious has the innate knack of offering up prose in story form. They *can* write automatically, and so they think that's the nature of writing itself, rather than *their* nature.

This explains E. L. Doctorow's famous assertion that "Writing is like driving a car at night. You can only see as far as your headlights, but you can make the whole trip that way."[1] Sure—*he* can make the whole trip like that and end up writing *Ragtime.* But when the rest of us follow suit, our stories almost always end up taking a meandering, disjointed, episodic route that often ends abruptly when we inadvertently drive off a dimly lit cliff. What's more, having a basic sense of story isn't always enough: sometimes a writer's debut novel is fabulous, but since they have no idea what it really was that captivated readers, their second, third, and fourth fall flat.

The belief that a story will appear if you write blindly into the darkness is extremely damaging, because it has given rise to one of the most seductive, widespread, and undermining concepts in the writing world: pantsing.

The Myth of Pantsing

There is a school of writing that believes the best way to write is to sit down, clear your mind, and write by the seat of your pants. Hence the term: "pantsing." (Yes, I know there's another meaning out there for pantsing; this is the writing one.) In some circles it's been dubbed the most authentic way to write. Part of what makes it so alluring is that it seems easy, straight-forward, and pure. Just let 'er rip! The goal is to let it all pour out as a way of "discovering" the story you're fated to tell. Knowing anything about your story before you begin to write is frowned upon as a surefire way to stifle your creativity and annoy the muse. Hey, as Robert Frost said, "No surprise in the writer, no surprise in the reader." It's a dubious sentiment at best, and often taken to extremes that would no doubt make Mr. Frost cringe, until it sounds a bit like Kevin Costner's mantra in *Field of Dreams*: "If you build it, they will come." Writing translation: write blindly and the story will magically appear. Instead, the surprise in both the writer *and* the reader tends to be the same: "Well, I thought this would be engaging; instead, it's a big fat mess."

But if pantsing leads to failure, why is it so damn seductive? Why *are* we so tempted to sit down and "let it all pour out"? Simple: We're hard-wired to do what's easy. This is not a negative. It doesn't make us weak, lazy, or slackers. It's just that thinking hard takes a whole lot of energy—after all, the brain accounts for only 2 percent of the body's volume, yet it consumes 20 percent of its energy. Thinking actually burns calories. (Not *enough* calories, but still.) So the urge to wing it is a built-in survival mechanism, the better to conserve precious energy for handling the decid-edly unexpected, the truly dangerous, the unavoidably challenging—you know, all the things that stories are about.

And let's face it, in the beginning it's much easier—liberating, almost—to let 'er rip and write forward without a second thought. Plus, since staring at that blank page can be exceedingly stressful, the relief of letting it all pour out not only feels good, it feels *right*. Thus it's easy to believe that this is, indeed, the natural path to the best-seller list. That is, until the exhilaration of winging it wears off, which is when writers often find themselves stranded on page 32 or 127 or 327. Or, just as often, page 3.

Chances are you know that feeling. You've been writing forward, cranking out page after page, and then suddenly you're lost. It's like you're standing in the middle of a big empty field, with no idea what comes next, or what matters, or where the story is going. And you think: *This is my fault, I'm a bad writer. Good writers automatically know what happens next. But when I shine my headlights into the darkness, all I see is fog.* Rest assured it's *not* that you're a bad writer. It's that you haven't been taught how to develop a story.

This brings us to the final, oft-cited myth about the glory of pantsing: that only by completely unleashing your creativity can you create a compelling story. Here's the thing: creativity needs context. It *needs* a leash.

Context is what bestows meaning and defines what matters, what doesn't, and why. Think of it as the yardstick that readers use to gauge the significance of everything that happens. After all, a rose is never just a rose. It's a sign of love from the cute guy next door—which makes it a wonder to behold. It's a wilted last-minute gift from your husband on your tenth anniversary—which makes it a major disappointment. It's what you were supposed to bring your girlfriend, but forgot, so it's the reason she broke up with you—which makes it a thorny reminder of your flaws. What do all these examples have in common? The rose got its meaning based on things that happened in the past, *before* it was presented (or not) to the recipient. The past determines the present. And when you pants, there *is* no past. Without the past to provide context, that rose is just a plain old pretty flower, and who cares about that? In a novel, the past—the things that happened before it began—are what provide context. For natural storytellers like Doctorow, whose cognitive unconscious *did* come equipped with a built-in understanding of story, the past and the present emerge together, on their own, in one fell swoop. If you or I had that innate ability, we'd be best-selling authors ourselves.

So rather than unleashing your creativity, you want to harness it to the past from which your story arises. Without the past to anchor the present, everything will be neutral and nothing will add up, and so it will come across as random to the reader. The resulting manuscript, although it might have a few fabulous pages here and there, will be a complete do-over.

The Myth of the Shitty First Draft

Wait, you may be thinking, *isn't that par for the course? Rough drafts are supposed to be shitty; Ernest Hemingway said so.* And he's absolutely right. But it's easy to misunderstand exactly what he means, a mistake even the otherwise brilliant Anne Lamott makes. She wholeheartedly embraces the notion of "really, really shitty first drafts," but then she defines them as "the child's draft, where you let it all pour out and then let it romp all over the place, knowing that no one is going to see it and that you can shape it later."[2]

Whether we're talking about your blueprint, or the entire first draft of your novel, she couldn't be more wrong—someone *is* going to see it, and that someone is the most important person of all: you. And chances are, after months of pantsing what you'll see is a collection of events that don't add up to anything—just a sprawling, aimless frolic. And trying to shape it only makes it worse, because there's nothing to shape. What's more, you've grown so attached to every thing in it, that editing, cutting, or rewriting feels a bit like sacrilege. So you massage it a bit, moving things around here and there, hoping that'll do the trick. It won't. The very fact that you *can* move things around is a telltale sign that the novel has no internal logic.

Will your first draft be shitty no matter what? Probably. It's kind of a badge of honor. But make no mistake: there's a massive difference between the shitty draft of an actual *story* and a shitty first draft that randomly romps all over the damn place.

The good news is that there's another school of writing that takes the opposite approach. The bad news is that it does just as much damage.

The Myth of Plotting

Adherents to this school of writing are known as plotters. Their philosophy is that the first thing a writer needs to do is outline the plot—that is, the novel's surface events—before they write one word, so they know exactly what's going to happen, from "once upon a time" to "and they lived happily (or not) ever after." Plotters come *so* close to being right. Developing a blueprint of the novel you're writing before you tackle page one is essential. The trouble is, they've focused on developing the wrong thing—the external plot—rather than the internal story. Their focus is on the external "what," rather than the protagonist's preexisting, internal "why."

Thus plotters begin by laying out the surface events of the story—beginning on page one—with little regard to the protagonist's specific past, which is the very thing that determines not only what will happen in the plot, but how she sees her world, what she does, and most importantly, why. Plotters have it backward: the events in the plot must be created to force the protagonist to make a *specific* really hard internal change. And that means you need to know, *specifically*, what that internal change will be before you begin creating a plot. Outlining the plot first is like saying, "I'm going to write about the most difficult, life-altering series of events in the life of someone whom I know absolutely nothing about."

That is as impossible as it sounds, short of actually channeling a muse who whispers the entire story in your ear as if she were reading it straight from a finished book. And speaking of finished books, that is exactly where the next writing myth stems from—not *a* finished book, but many of them, not to mention movies, plays, and *actual* myths.

The Myth of External Story Structure Models

Outlining the plot before you develop your protagonist traps you on the surface of your novel—that is, in the external events that happen. Without the protagonist's preexisting inner story to guide them, writers often feel a tad lost. Since all the tension, conflict, and drama must therefore come from the plot alone, how *do* you decide what happens and in what order?

To help writers in this pursuit, there are numerous guides that very deftly outline what has been dubbed "story structure." The problem is, story structure is a misnomer because these guides are not about story at all. They're about plot structure, which is very different indeed. It all began with one of the most beloved "story" structure models, Joseph Campbell's hallowed *Hero's Journey*. Its basic premise, and that of all the structure guides that followed, is that a lot of myths, novels, and movies have a similar structure and shape. Ironically, although these guides often hint at said hero's internal struggle and are full of words like "quest" and "challenge," what that struggle might actually *be* is never discussed, nor is how you'd figure it out, or the role it plays in creating the plot itself. Instead, these guides zero in on the sequencing of events in and of themselves, as if each "hero" gets tossed into a one-size-fits-all gauntlet. So something "big" happens on page 20, something "dangerous" by page 50, something definitively "upsetting" by page 100, and so on. Successful stories often *do* follow the external pattern these guides set forth, so it's deceptively easy to believe that all you have to do is ape the shape and you've got a story. I've read a surprising number of manuscripts that fall into this category. You can always spot them early on; they have a paint-by-numbers predictability—it's all about form, with very little substance.

Here's the thing: every one of those models is based on tons of *finished* novels, myths, or movies. Do you think the *creators* of those original myths were cribbing from external story structure models when they came up with their tales? Sheesh, it would've been kind of difficult, since many of them originated long before there was written language, let alone a slew of handy guidebooks. They told stories built on content, not on structure. Writing a novel based on such an outline is like painstakingly assembling one of those exquisitely rendered models of the human heart and then expecting it to start beating all on its own. If only!

Story structure is actually born of a story well told. It's not something you can create from the outside in. Trying to write a novel based on an external story structure model often leads to genuine heartbreak for writers. They follow the instructions religiously, and sure enough, the surface shape of their novel matches the prescribed structure, beat for beat. And yet, it's not nearly as engaging as all those novels, movies, and myths that the "story structure model" was based on.

At that point it's all too easy to believe that what's really missing is simply a bit of spit and polish. So rather than diving into the story and rewriting it from the inside out—which almost always means a do-over starting on page one—the writer then eagerly buffs it from the outside in. Ironically, this tends to highlight the novel's problems rather than remedy them. Before, the emphasis was on the external events; now it's on the pretty language that describes them. Rather than inviting us in, the beautiful language is more like a waterproof sealant, locking us on the surface where all we can do is admire the words, rather than absorb the story that they're meant to tell. And while this is the last thing the author meant to do, now we're kind of annoyed. Because it feels a wee bit like the writer is showing off. Like he's saying, "Look how well I write!" rather than "Forget about *me*; lose yourself in my story!"

But if these three techniques—pantsing, plotting, and following external "story" structure models—don't work, what does? The solution springs from what we've learned about story's effect on the brain, and it starts, proceeds, and finishes with your protagonist's inner struggle—your story's third rail. Creating the inside story comes first, because without it you can't create your plot. Let's take a look at what this actually means.

You Can't Have an After Without a Before

At its most basic, a story is about how someone grapples with a problem they can't avoid, and how they change in the process. Understanding that truth will make the difference between spending years writing a highly polished but ultimately dull rendition of "things that happen" and a riveting novel that readers won't be able to put down.

Here's the skinny: You can't write about how someone changes unless you know, specifically, what they're changing *from*. You can't write about a problem unless you know, specifically, what caused it. And as real life has taught us all too well, by the time we're forced to face a thorny problem, chances are it's been building for quite a while—years, decades, often our whole life up to that moment.

The story you're telling doesn't start on page one. It started long before you got there.

Every novel begins *in medias res*, a Latin term meaning "in the middle of the thing." As Horace pointed out three millennia ago in *Ars Poetica*, it's far better to begin in the middle of things rather than *ab ovo*—a fancy way of saying "from the egg." He was praising the work of Homer's *The Iliad* when he said, "Nor does he begin the Trojan War from the egg; but always he hurries into action, and snatches the listener into the middle of things . . ." [3]

Make no mistake, *in medias res* is not a literary device. It is a fact. Unless, of course, you *are* going to begin your novel with "Achilles, the son of the nymph Thetis and Peleus, the king of the Myrmidons, was born on a cool spring morn . . ."—which, I'm guessing, you aren't.

The trouble is that writers often misunderstand the notion of what beginning *in medias res* means, taking it to simply be a device to plunge us into current action and explain it later. Not only is that not what it means, but to do so is a tragic mistake. Because by leaving the "why" out of the picture, the action often reads as a bunch of things that happen. Worse yet, writers are often so focused on getting the "what" onto the page that they, themselves, don't even *know* the "why."

Here's the real truth: your novel itself begins "in the middle of the thing"—the "thing" being the *story*. What starts on page one is the *second* half of the story, when the plot kicks in. The second half—the novel itself—will contain large parts of the first in the form of flashbacks, dialogue, and snippets of memory as the protagonist struggles to make sense of what's happening, and what to do about it. It bears repeating: nothing in this process goes to waste.

But the simple fact remains that without the first half of the story, there can be no second half. The first half establishes where the problem came from and who the protagonist is to begin with, so that the plot you then create can force her to struggle with that problem and, in the process, change.

That's why we're going to spend the next several chapters unearthing the first half, which is what will allow you to pinpoint exactly where your novel begins. Only then can you begin your novel's official blueprint, as you *continue* creating your protagonist's story and begin thinking about the plot. The beauty is that by then the plot itself will have already begun to appear on its own because the past gives birth to the future.

It's no surprise that Dante began *The Divine Comedy* with

> Midway in our life's journey, I went astray
> from the straight road and woke up to find myself
> alone in a dark wood.[4]

With that in mind, into the woods we go . . .

PART II

CREATING THE INSIDE STORY

3

THE *WHAT IF?* EXPECTATION, BROKEN!

No idea is so outlandish that it should not be considered.
—WINSTON CHURCHILL

Here's something we probably don't need neuroscientists to tell us: the brain craves certainty. We like to know things for sure, so we can plan accordingly. I mean, we have an entire channel devoted solely to the weather, so we'll know if we should take a sweater when we head to the market.

But let's face it, no one knows what the future will bring, not even Al Roker on his best day. Sure, we expect certain things to happen—taxes will rise, Uncle Howard will ask embarrassing questions at Thanksgiving, and the Yankees will never have another season like '98. But because we can't be 100 percent positive, we tend to spend a lot of time wondering *what if?* Yes, we're dreamers, but with a purpose—we want to be able to survive the unexpected problems that the future might have in store for us. After all, the one thing we *do* know with absolute certainty is that the unexpected often happens, in spades, and we want to be prepared when it does—you know, so we don't make fools of ourselves, and wear flip flops and shorts to a blizzard.

Story evolved as a way to envision the future and thus plan for the unexpected—lest those alternate scenarios catch us completely unaware.

That's why we'll always need stories—because the unexpected keeps being so darn unexpected—and it's why we'll always need writers to explore those *What If* scenarios for us.

Every story, from the simplest picture book to the most epic fantasy novel, begins with the question: *Hmmm, what if?* It's not that we necessarily ask this question consciously. Rather, story ideas tend to start with something out of the ordinary that strikes your imagination like a pebble tossed against your window at midnight, but that can feel like a rock smashing it to smithereens. The idea that sparks a novel can be the briefest snippet, the most gossamer notion, a single arresting image, yet it has the power to take hold and catapult you out of the familiar world of *What Is* and into the intriguing, unexplored world of *What If.*

The first step is to transform that initial slip of an idea into a potent *What If* question. That sounds kind of simple, doesn't it? As if all you have to do is envision something out of the ordinary, plunk the words *What If* in front of it, and you're ready to start writing. In fact, that's exactly what we've been taught to do, beginning in kindergarten. The only problem is, it doesn't work. Turns out there's a wee bit more to a story-worthy *What If* than simply imagining a decidedly unexpected turn of events.

In this chapter, we'll explore the danger of a neutral *What If*; we'll delve into why a *What If* must revolve around something unexpected that throws a monkey wrench into someone's well-laid plans; we'll learn that behind every successful *What If* is something even more seminal: the point you want it to make. Finally, we'll learn how to create a potent *What If* from the merest wisp of an idea before it evaporates.

What Kindergarten Got (and Still Gets) Really, Really Wrong

Over the years, I've read countless heartbreakingly storyless manuscripts that dedicated authors spent years earnestly sweating over, but that still somehow turned out to be nothing more than a bunch of big, eventful, unusual things that happen. These are the same 96 percent of manuscripts that agents reject out of hand. Given the confident cover letters accompanying these manuscripts, it was clear that the authors were pretty sure that what they'd written was not only a story, but a really compelling one at that. They weren't fools—they were smart, knowledgeable people—and for a long time I wondered why so many otherwise competent writers went so wrong in the exact same way. Where did this mistaken notion about story come from? I stumbled on a big part of the answer last year, when I had the privilege of helping a small maverick school district in New Jersey incorporate story into their writing program.

What I learned is that the notion that a story is "a bunch of big, eventful, unusual things that happen" is firmly planted in kindergarten and nourished from there on—which is why it can be so hard to uproot. It's at the foundation of how narrative writing is taught, and a major reason why so many manuscripts fail. So let's revisit elementary school for a glimpse of how our well-meaning teachers may have accidentally planted beliefs that have hampered our writing ever since, and take a look at the brain science behind why those beliefs feel so true.

Okay, grab a juice pack, sit back, and imagine you're in the third grade—hey, pay attention, no passing notes! Now, your teacher tells you it's time to do some creative writing. You pull out your notebook, pick up your pencil, and wait for her to give you the prompt, the *What If* question that you'll be writing about. The prompt will be exactly like these, which are shortened versions of actual *What If*s culled from statewide testing in New Jersey:

- What if Jane was walking along the beach and she found a bottle with a message in it? Write a story about what would happen next . . .
- What if Freddy woke up and discovered that there's a castle in his backyard? He hears a strange sound coming from inside, and then . . .
- What if Martha walks into class and finds a great big sparkly box on her desk? She opens it and inside she finds . . .

Sure enough, all these *What If*s posit an unusual, often dramatic, unexpected event. Each event flies in the face of what the characters *thought* would happen, forcing them to confront something decidedly out of the ordinary—so far, so good. It feels like the start of a story, doesn't it? Because that is, in fact, what the brain is always on the lookout for: something different, something unusual, something that we should pay attention to now, lest it turn out to be the sort of problem that will clobber us the second our back is turned. That's why another word for "the unexpected" is *surprise*!

Surprises make us curious, which is why these prompts seem like the perfect place to start a story. The problem is, these surprises don't lead anywhere, because they lack the essential element we were talking about earlier: context. Let's examine why it's so easy to miss that fact when confronted with a storyless prompt.

You Had Me at *Surprise!*

Surprise instantly commands our attention precisely because it defies our expectations. Once engaged, we're wired to immediately start figuring out what's actually going on, the better to gauge whether we're about to get whacked or kissed. This "sit up and take notice" response to surprise is a survival mechanism; without it, we'd blithely ignore every warning sign ever thrown at us, because we'd have no idea that anything out of the ordinary was going on. Instead, surprise knocks us out of our dependable routine and forces us to consider as yet unexplored realities: You know you parked your car in front of your house, but it's not there now. You expect your significant other to be home by six, and now it's midnight. You were sure your dog was housebroken . . . uh-oh. The unexpected is surprising because it (1) challenges what we *thought* was going to happen and (2) puts a have-to-deal-with-it-now crimp in our well-laid plans, even if those plans were simply to head home after work, order takeout, and watch *Sleepless in Seattle* for the thousandth time.

A broken pattern forces you to reconsider something that, up to that moment, you tacitly assumed you could count on. That's how the brain rolls. We have what scientists call an "avidity for patterns." That's why from birth on we're constantly scanning the terrain for reliable patterns. *If I cry real loud, that nice lady will feed me. Got it!* Patterns translate what would otherwise seem terrifyingly random and chaotic into a reassuringly reliable order we can make sense of. Think: *If this, then that.* It doesn't mean we necessarily like the pattern we see, mind you; it just means we know how to safely navigate it, which might explain why we tend to stick with the devil we know.

Once we recognize something as a pattern, we tell ourselves a story about how it works—explaining the "why" behind it, and therefore how we should handle it. And then we forget all about it. In our conscious brain, that is. Not because our memory is faulty, but because we've tucked the info into our trusty cognitive unconscious—the part of our brain that makes almost all of the thirty-five thousand decisions we face each day—freeing our conscious mind to be on the constant lookout for, yep, the crafty old unexpected. We don't tend to give a second thought to the things we've come to expect.

That's why I'm betting that none of you went to bed last night thinking, *Boy, I sure do hope the sun comes up tomorrow, because I have a lot to do and if it didn't, that would definitely slow me down.* But imagine what would happen if the sun *didn't* rise. Talk about a familiar pattern, broken! That's exactly the sort of thing that we could write a story about, because navigating a situation like that is definitely out of our normal, everyday wheelhouse.

Thus it makes total sense that a *What If* revolves around a pattern-breaking surprise. So what's wrong with simply speculating, "*What if* the sun doesn't come up?" and starting to write a story? A lot, as it turns out.

What Ifs That Don't Live Up to Their True Potential

There's no denying that the creators of the *What If* writing prompts we've been getting since kindergarten were onto something. The initial snippet of a story idea *does* begin with a surprise, something that upends our expectations. But there's a whole lot more to it than that.

Because *so what* if Freddy discovers a castle or Martha finds a big box on her desk or Jane finds a message in a bottle? Unless we know why these things would matter to Freddy, Martha, or Jane, they're just a bunch of unusual things that happen, even if they *do* break a well-known external pattern. Not only don't they suggest an actual story, they don't suggest anything at all, other than the reaction: *Wow, that's weird!*

*What If*s like that are debilitating to writers, whether you're eight or eighty, because they're utterly random and thus completely neutral, rendering them pointless. You could use them as a starting point to write absolutely *anything*. Freddy finds out he's a king! Or a wizard! Or a boy whose toys came to life! That should be liberating, right? You can unleash your creativity and see where it takes you.

But here's a counterintuitive fact: the prospect of endless possibility isn't freeing, it's paralyzing. Myriad studies have shown that the more choices we have, the less likely we are to choose *anything*. Not only that, but limitless choice tends to trigger anxiety. How do you know what the right choice is when you have a gazillion? What if you pick the wrong one? Plus with so many to choose from, no matter which one you pick, the first time you hit a snag that smug little voice we all have pipes up and says, "See, I *told* you that you should have picked the other one!"

That's why children often freeze in the face of such writing prompts. And unfortunately those poor little kids don't have the advantages that we adults do when faced with such endless possibility—like the ability to suddenly remember that we absolutely have to, um, oh yeah, clean out the refrigerator. Nope, with the teacher watching, those kids *have* to write something. And that something usually turns out to be a string of equally dramatic, random, pointless events that sound something like this:

Freddy went into the castle and found a clown from Mars, who had landed in a spaceship shaped like a unicorn, Freddie leapt inside and flew back in time to the Middle Ages, where a knight was battling a giant octopus with purple tentacles and, and, and then Freddy woke up and discovered it was a dream. The End.

You see what I mean? Those prompts offer no help at all when it comes to writing a real story, which is probably why so many of the kids' stories end up being a dream. Even at the tender age of eight, we know when we've dug ourselves into a narrative hole so deep that the only way out is to do what the writers of the TV series *Dallas* did when they realized that an entire year's worth of episodes were just a bunch of things that happened: declare it a dream and move on.

Sadly, this misconception about the nature of a potent *What If* doesn't end once we leave elementary school. Want proof? Here are three writing prompts that sure do sound like the *What If*s those kids were struggling with. They were culled from one of the nation's most popular creative writing sites. They've been paraphrased for space, but the gist remains the same:

- What if a wizard's terror bolt lances overhead? And so, dagger held high and ice shard at the ready, you tear toward the dastardly spell caster, and . . . ?
- What if you're awakened at midnight by your dog barking wildly and you look out the window, only to see a face staring right back at you?
- What if a fortune teller at the local county fair tells you two things: something good that will happen, and something awful that will happen?

The trouble with these *What If*s is that although something odd and externally dramatic happens in each one, an unexpected problem alone cannot drive a story. By itself, it becomes nothing more than an in-the-moment external problem that, at most, demands a bit of surface derring-do to solve. This is because, inevitably, the plot will focus solely on the strange event, rather than the effect said event might have on a specific person. They are just—say it with me—a series of big, eventful, unusual things that happen. And that, my friends, is how we end up with novels that go nowhere.

What's Your Point? Or How to Turn Any *What If* into the Start of a Story

The reason it's so important to debunk what your elementary school teachers, in good faith, taught you, is because it most likely took root and has been affecting how you've seen writing ever since. Hell, you have only to glance at the prompts that came from that popular writing site to see how deeply ingrained the notion of a context-free prompt is in our collective writing conscience. It's damn hard to unlearn something you've inadvertently internalized since you were five.

That's exactly what we wanted to save those kids in New Jersey from, while they were still blissfully drawing with every crayon in the box. We didn't want to leave them at the mercy of problemless prompts, especially with state-mandated testing looming. What the kids needed was a concrete strategy so they could walk into that test feeling confident, knowing they could turn any neutral *What If* into the springboard to actual story. So I devised a way for them to bring an internal problem to the external prompt. How? By having them ask the question that adult writers need to begin asking the moment that first intoxicating glimmer of an idea strikes.

What's your point?

It's easy for writers of all ages to lose sight of one very simple, grounding truth: all stories make a point, beginning on page one. Which means that as a writer you need to know what that point is, long before you *get* to page one. Especially since the point your story will make is what allows you to pinpoint the one thing all those surface *What Ifs* are missing: the source of your protagonist's internal conflict. In other words, the very heart of the story, and what it's really about. That's why the first thing you need to do, we told the kids, is to decide what point you want your story to make, because the point will tell you exactly what kind of internal problem your story will be about. For instance:

- Friends stick together when times are tough.
- Believe in yourself even when others don't.
- Think about how others will feel before you act.

All by themselves each of these very simple points suggests a potent internal story problem. To wit:

- A group of friends will face a tough problem sure to challenge their loyalty to each other, ultimately teaching them the hard-won benefits of sticking together (or not). Think *Sisterhood of the Traveling Pants*.
- A character will want to tackle something tough that she's never done before, and when everyone tries to convince her that she can't do it, instead of giving up, she'll muster the inner courage to give it her all (or not). Think *Mulan*.
- A character will *really, really* want to do something that, okay, might hurt someone else, and will have to struggle with whether to do it (or not). Think *Indecent Proposal*.

Each of these points gives us a glimpse of the hard internal choice that those big, externally dramatic *What If*s will force the protagonist to confront. Here's the key:

1. The point is what is borne out in the protagonist's inner struggle.
2. The *What If* centers on the external plot that will trigger that struggle, ultimately making the point.

The point is what transforms a neutral *What If* into one with the power to begin bringing a story to life—and the operative word here is *begin*. Because the other mistake your elementary school teacher made was thinking that once you have your *What If* it's time to start writing that story. Not so.

While a *What If* harnessed to a clear point contains the initial elements of a compelling story, it's only the first step in taking that tantalizing wisp of an idea and making it something concrete. Since the story itself is born of the specifics, no *What If*, regardless how perfectly stated, could uncover those specifics. So don't worry if the *What If* you're about to write is messy and not perfect. There is a big difference between an exploratory, fledgling *What If* and a *What If* that sums up a novel that's already been written. In fact, your *What If* might sound simple, familiar, and rather flimsy. That's fine. Don't think of it as the basis of an entire novel, but as the first glimmer of concrete possibility. A working *What If* provides a great big "X marks the spot," so you know exactly where to begin digging for the specifics that will bring your story to life and make the point you want to make. For instance, let's say your point is: holding a grudge can have tragic unforeseen consequences. Here's a *What If* created to make that point with a vengeance (and yes, it's culled from a very famous tale of woe): What if two teenagers fell madly in love, only to discover that their parents are mortal enemies? (*Romeo and Juliet*.)

What's so great about this *What If*? The surprise that it's built around ("Your last name is *what*? Uh oh!") is personal, and guaranteed to shatter those starry-eyed teens' dreams, leading to inescapable conflict as they struggle to avoid an impending consequence that matters to them more than anything else in the world (including, apparently, life itself). The key element is this: the inherent, unavoidable external conflict implied in that *What If* (their parents' long-standing feud) is going to trigger inherent, unavoidable internal conflict (their intense desire to be together, despite their family loyalty). You can instantly envision scads of potential escalating action based on that one ongoing problem.

One note: R&J is a classic story, one we all know. So in hindsight it's easy to pluck out a short, compelling *What If*, one that not only captures the essence of the story but also implies the point it will make—especially because we *already know it*. I use it because it's a clear, concise example of what you need to know about your story in order to begin, *not* because your *What If* needs to mirror it. Plus, your *What If* isn't for anyone but you. And it doesn't have to be just one line. It might be three or four lines, even five. Your goal is simply to make sure you have a point that will bring your *What If* to life.

So Shakespeare aside, what does *that* look like when a writer is just starting out on a project? It's one thing to deconstruct something that's already been written, and quite another to watch something being developed from the ground up. That's exactly what we're going to do, by following one case study throughout the book.

I knew that my friend Jennie Nash, novelist and book coach, had been toying with an idea that, at the moment, was nothing more than a tiny glimmer. So I asked if she'd be willing to pinch-hit for us and develop it here, step by step, and she said yes.

Before we begin, a word about what to expect from Jennie's examples. We're used to reading finished work that's so polished and seamless it's hard to believe there was a time when it was in development—a fledgling thing not yet ready to fly. In fact, it can be hard to shake the notion that all the novelist did was sit down, bang out "Once upon a time," and keep on typing until she reached "The End." After which she immediately emailed the manuscript to her publisher, and the next day you downloaded it on iTunes.

That image is often what stops promising new writers from getting past the first *sentence.* They're sure that a story not only comes to you whole, but wrapped in exquisitely worded sentences. So when they can't even come up with one relatively perfect opening sentence, they take it as an omen. And give up. Don't! That's the literary equivalent of marveling at an exquisite bone china vase—so delicate, so translucent, so refined—and imagining that it sprang into the world in perfect form all of a piece, instead of being created out of a sticky mixture of crushed, calcinated cow bone ash, pulverized china stone, and kaolin clay. But while you have no fantasies about the origin of Aunt Milly's prize vase, it's hard to shake that illusion when it comes to novels.

Part of what we are doing here is blowing the lid off that myth. We're getting a peek beneath the glossy cover, watching as a working writer begins to create the material from which to spin her tale. Some of what you'll see may well end up in the pages of Jennie's novel; in other places it will be undeveloped and rough. What you're watching is the gritty evolution of the first scrap of an unformed idea into the basis of a compelling novel. Seeing the process unfold isn't just revealing, it's comforting. Because not only doesn't *any* of it have to be perfect out of the starting gate, it can't be. It's the same as if you were, in fact, making a bone china vase: first you have to dig up the ingredients, then you have to refine them, mix them together, and finally, once you've created the clay from which to build your novel, begin to bring it to life.

One last note: Story is story, regardless of the format and genre, but we all have our favorites—the type of novel we lean toward, the genre that speaks to us. Jennie writes contemporary fiction. Although that might not be your genre of choice, the steps she will go through to develop her blueprint apply across the board. While there are some slight differences if you write thrillers, mystery, police procedurals, or speculative fiction, we'll point them out along the way.

So sharpen your pencils, ink up your pens, dust off your laptop: it's time to begin!

Step 1: That First Pinprick

Shakespeare probably knew that after their night of unparalleled bliss, Romeo and Juliet weren't going to survive much longer on this earthly plane, which tipped him off that daggers, swords, and poison might come into play. You may know all kinds of specific things about your story, too. You may even have a chunk of chapters written, or a whole first draft of your novel. Let's ignore all that for a moment. Let's go all the way back to the beginning, to that first moment when that wisp of an idea struck you. Take a deep breath, feel it in your bones. Close your eyes even. Now, can you zero in on the instant that you first felt the pinprick of the idea?

It might be a single surprising image that caught your attention. That's what happened to E. L. Konigsberg, author of *From the Mixed-Up Files of Mrs. Basil E. Frankweiler*, a novel about a young sister and brother who run away and hide out for a week in the Metropolitan Museum of Art in New York. According to Konigsberg, the first snippet of the idea came like this: "My three children and I were visiting the museum, wandering through the period rooms on the first floor, when I spotted a single piece of popcorn on the seat of a blue silk chair. There was a velvet rope across the doorway of the room. How had that lonely piece of popcorn arrived on the seat of that blue silk chair?"[1]

Or that first pinprick might revolve around a point you want to make. Stephen King has said of *Under the Dome* that "From the very beginning I saw it as a chance to write about the serious ecological problems that we face in the world today."[2]

Or it might have been an actual *What If* question that first grabbed you. Like: *What If* a lawyer couldn't tell even the teeniest, tiniest lie? (*Liar, Liar*). *What If* you could see what the world would be like if you hadn't been born? (*It's a Wonderful Life*).

Or as we'll see with Jennie's answer, it might be a sudden unanswered question that pops up and won't let go. Here's what Jennie said: "I kept thinking about a story with a woman at the center who doesn't like dogs. That's all I had—this woman with this strange and somewhat unpopular characteristic. My friends happen to know that even though I had three wonderful little miniature Schnauzers as a child, I don't actually like dogs. They kept asking me about that—'What don't you like about dogs?' I didn't have a clue, but because of this story idea that kept pinging in my head, I began to think more about it.

"I thought about all the times my kids asked me for a puppy and how I always said no. There was no way I was going to get them a dog, because I knew two things: how much they'd love the dog, and that sometime in the future that dog would die. Which meant that I'd have to get both the girls (and myself) through the loss, and I did not want to go there. Ever. I once said this out loud to my grown children, and they couldn't believe that this crazy irrational belief was what kept them from having a dog. They were astonished and mad! I knew it was a crazy irrational belief, but I believed it all the same.

"That's when I started thinking more deeply about a character in a story who didn't like dogs." Here comes the pinprick that turned this from a random collection of memories and thoughts in Jennie's head into the seeds of something more: "What would happen, I began to wonder, if a woman who didn't like dogs was afraid to love—dogs, yes, but everything and everyone? What if she was convinced that the loss she would experience simply wasn't worth it, and so she lived her whole life with this mistaken belief? And what if something happened (with a dog—it was always going to be with a dog) that then proved to her—too late—that she was wrong?"

That was it—the tiny turn of the screw that led Jennie to her story.

WHAT TO DO

Now it's your turn. In no more than a page, write about the instant the idea that you're working with—the one that won't seem to leave you alone— first grabbed you. As Jennie did with her fledgling protagonist who doesn't like dogs, try to zero in on the very first glimmer.

Step 2: Why *Do* You Care?

Now the question is, why have you even gone this far with it? Ask yourself, *why does this stick with me? Why do I care about it?* The goal is to burrow into what it is about this idea that intrigues you. Don't worry if your answer is still kind of fuzzy, or if your first thought is, *Um, now that I think about it, what really strikes me is that I need a nap.* There is no right answer. You might even find that what you originally thought drew you to the idea was wrong, and that it's something else altogether. It's all exploration at this point. Your goal with all these questions is to zero in on the heart of the story you *want* to tell.

Note that this process—even right here at the start—can be scary. Because you really do have to ask yourself, *what* does *matter to me?* The most exhilarating part of this process is that *this is often how you find out what things mean to you.* As Joan Didion so famously said, "I write entirely to find out what I'm thinking, what I'm looking at, what I see, and what it means. What I want and what I fear."[3] And that's what all this digging is about. Because as writers, there is one thing we all want: to communicate. To reach other people. To be part of the great conversation of the human race. That, and maybe make a little money, too. After all, a writer has to eat.

Jennie cared about her question because the notion of grief has always fascinated her—and not for any obvious reason like she lost her father as a child or was traumatized by a screening of *Bambi*. It's something she has often thought about: "How close love sits to death, how you can't have love without pain." Mixed in with the idea of grief, for her, is the idea of regret—of *not* loving because of the fear of what might happen. "Some of my favorite novels are about regret and loss," Jennie says, like *The Remains of the Day* by Kazuo Ishiguro, *The Soloist* by Mark Salzman, and *Girl in Hyacinth Blue* by Susan Vreeland. The question she was intrigued by suggested this kind of story, and so it began to come to life in her mind.

WHAT TO DO

In no more than a page, write down why you care about the story that you want to tell. There is no right answer; whatever comes to mind is relevant—even if it seems silly. You might surprise yourself and discover you care for an entirely different reason than you thought you did.

Step 3: What Is Your Point?

The real question here is, what do you want your readers to go away thinking about? What are you trying to say about human nature that will help us keep from getting trounced in the future? Sure, it might shift, expand, or even double back on itself as you continue exploring your story, but you can't write forward without having a point. It's what starts focusing you on the specific problem the story will tackle, and even more importantly, on what that problem will mean to your protagonist. Your starter answer may be general and as simple as *kindness matters most* or *we are more resilient than we realize* or *technology has rendered us helpless in ways we don't yet understand*. The goal isn't to nail the specifics now, it's to let you know where to look for them. Because in this as in all things, the deeper you dig, the more grist you have for the mill.

What's Jennie's point? "Well," she said, "in my original idea, there was that notion about the fear of loss keeping us from experiencing love. So my point would be, perhaps, the opposite—that love is worth what it costs. That it's better to have loved and lost than never to have loved at all."

Oh no, you might think, *that's a cliché! They'd laugh me out of my writer's group.* In fact, writers often secretly confess that their biggest fear is that what they're writing about is so common, so small, that no one will be interested. Ironically, that is exactly what people are interested in. Why? Because those common, everyday things like love, loyalty, and trust are things we all experience, and we're always looking for new insights that might help us navigate our everyday lives in a new and fresh way. While none of us will ever be a robot on Mars battling intergalactic warriors (I don't think), just about all of us have a tricky relationship with someone that we'd love to have help figuring out. Which is why an effective novel about a robot on Mars would probably center on love, loyalty, and trust, too. As for Jennie's point, believing that love is worth the cost is actually pretty crucial when it comes to survival—social survival, as well as the actual survival of the species—since love so often leads to sex, which so often leads to the next generation.

Here's the (very reassuring) skinny: at the beginning, just about *every* story starts with a cliché. A cliché is simply something that's so familiar that it feels old hat. It's the story's job to make it, um, new hat. As Samuel Johnson so aptly pointed out, "The two most engaging powers of an author are to make new things familiar, and familiar things new."[4]

So don't be surprised if at the outset your point, like Jennie's, sounds mundane. What she has is a good start, and it already suggests where to dig a tad deeper—there is a story forming that has to do with deep loss, big regret, a choice someone made to protect themselves from something that could have gone very differently.

WHAT TO DO

See if you can nail the point your story will make in a few concise lines. Don't worry if in the beginning it splashes all over the page. Just keep focusing in on the single driving point it will make. The goal is to reduce it to its essence.

Step 4: Drafting Your *What If*

Fueled by your passion for your specific idea, and armed with the point you want to make, let's work on your *What If.* The goal isn't to nail it perfectly in the first attempt. In fact, trying to do just that is the number one reason writers give up, largely because it's like trying to lay the foundation, build, paint, and sell a house all in one fell swoop. Instead, you end up buried under a ton of bricks. So, let's do a bit of foundation-laying, because yes: even your *What If* has layers that you can revise to make it better.

For instance, Jennie's first attempt at a *What If* was back there in her glimmer of an idea. When I asked her to refine it, she wrote this: "What if a woman who was afraid of commitment realizes—too late—the error of her ways?"

Is this a good *What If?* Not even close. Yes, there's a nice little problem and resolution in there, but it's far too abstract. Plus this *What If* has nothing to do with a dog, even though Jennie knows her novel will have a dog at the center. If Jennie were to start writing that story, she would end up with lots of vague, surface scenes, with a dog or two thrown in for good measure—and no story at all.

So, armed with the knowledge of what's missing, Jennie went back and wrote this version of her *What If*: "What if a woman who has spent her whole life hedging her bets against love (of people, of things, of dogs) is forced by grief into a relationship with a dog that proves to her—too late—the error of her ways?"

Much better. But it could be even tighter, because right now we don't know what exactly this woman is grieving for, and we have a lot of questions about this dog and what's going on between them. So Jennie went back in and—giving herself the freedom of more than one sentence—wrote this: "What if a woman who's spent her whole life believing she's successfully hedged her bets against love (of people, of things, of dogs) is on the verge of losing everything—the one person she's felt close to, her lifelong career, and her grasp on reality? Mad with grief, she has one chance to set things right, but first she must convince those around her that she's not suicidal. So she devises a scheme to steal a dog for an hour or two, believing that 'getting' a dog will reassure the people in her life (who are dog lovers) that she's back on the path to emotional stability. But when she can't get rid of the dog, she's forced to confront the fact that the very thing she spent her life avoiding—connection—is what makes the inevitable grief of loss endurable."

Bingo! That *What If* has context, a surprise, and conflict leading to consequences that will clearly spark the novel's third rail—the interior struggle that this character is grappling with. But what is that struggle, exactly? Many questions immediately pop up, not the least of which are these: Why can't she get rid of the dog? And "set things right"—what things? To whom? And who *is* she, anyway? At the moment Jennie has no clue, and this vagueness is absolutely fine. In fact, it's the point. The immediate goal of your *What If* isn't to tell strangers what your book will be about, or to hook readers. It's to tell you, *yourself*, what you'll need to discover before you can begin to craft your plot.

WHAT TO DO

Now, you try it. Write a *What If* that's as fully fleshed out as you can make it, but still concise. You don't want this *What If* to sprawl all over the place. Don't be afraid if the first attempt is wide of the mark. Keep at it until you have something specific, with context, conflict, and a hint of surprise. In essence: something that will make your point.

4

THE WHO: WHOSE LIFE WILL YOU UTTERLY UPEND?

I think the whole glory of writing lies in the fact that it forces us out of ourselves and into the lives of others.

—SHERWOOD ANDERSON

Now that you've got your *What If*, we're going to set it aside for a moment. I know, I know—you've got such a great situation set up, and it suggests so many intriguing possibilities that it's tempting to linger up top and noodle around with the plot, just to see where it takes you. I can't stop you. But remember, no one ever asked, "Whose plot is it?" The question you need to answer before you can develop the plot is the one writers *are* always asked: "Whose *story* is it?"

As we know, the story and the plot are two very different things. The story comes first, and it is born of one person, and one person only: the protagonist. Everyone and everything else will be created to serve his or her story. A novel's power depends on how deeply you dive into your protagonist—that's what will bring your plot into being and give it life. So rather than asking who will run through your novel's preordained gauntlet of challenge, the goal is to figure out *who* you'll build that gauntlet to test.

In this chapter we'll discuss why every story needs a protagonist; why it's crucial to zero in on your specific protagonist before you start plotting; how choosing a protagonist immediately begins to shape your *What If*; how to select the most compellingly conflicted, unsuspecting protagonist to make your point and carry your novel; why there is almost always a single protagonist; and how, when there seems to be more than one main character, to identify the alpha-protagonist.

The Importance of Playing Favorites

I remember the very first time I spoke to a writer's group as a story analyst. It was a small, intimate gathering, and the plan was that each writer would read the opening of her novel, and I'd give her feedback. I was as nervous as they were. What if I had no comment at all or, worse, said something dumb? Happily, that didn't happen; we quickly settled into a comfortable routine, and I began to relax. And then the last writer, the one who'd baked the delectable cookies we'd all been nibbling on, started to read.

Her opening pages introduced more characters than I could count, doing things with great gusto, but for no apparent reason that I could see. My heart began to pound. What question could I ask, since I had no idea what was happening, or why any of it mattered to anyone, or where it was going, or, well, anything? *Stall,* I thought. So when she finally finished reading and looked at me expectantly, I cleared my throat and asked the only question I could think of.

"I know *you* know the answer to this," I began, "and I may be a dolt, but I'm not quite sure who the main character is." I figured once she told me, maybe I'd have a handle on what all the crazy comings and goings were about.

She blinked, genuinely confused. The room went silent. My first thought was that they were all thinking that I *was* a dolt not to see that—Edna? James? Rachel? Fido?—was clearly the protagonist.

Then the writer finally said, "Gee, I don't know, I never really thought about it because I have so many characters. Do I really need a *main* character?"

The thing is, that writer wasn't a dolt, either. It's a very common mistake, one born of the understandable misbelief that a story is about the things that happen in it, which would definitely affect a whole lot of people. So why *is* it necessary to pick one person to be your protagonist?

Your Protagonist's Brain Is Your Reader's Portal

The answer brings us right back to why we're wired for story. The world is teeming with things that happen, and on most days, especially before that first cup of coffee, it sure looks like chaos out there. Our survival depends on making sense of the particular chaos we call home—not in the general "objective" sense we hear so much about, but in the much more practical, subjective, how-will-this-affect-me-personally sense.

Thus the evolutionary job of story is to funnel said chaos through one very grounding filter: the specific effect that chaos has on the protagonist, who becomes our avatar. The events by themselves mean nothing; it's what those events *mean* to someone that has us compulsively turning pages. That's why the protagonist is the portal through which we enter the novel. Remember, when we're lost in a story, we're not passively reading about something that's happening to someone else. We're actively experiencing it on a neural level as if it were happening to us. We are—literally—making the protagonist's experience our own.

Without a main character, the reader has no skin in the game, and everything remains utterly neutral and surprisingly hard to follow. While we might know *what* is happening, we have no idea why it matters or what the point is. Because the point doesn't stem from the events; rather, it stems from the struggle they trigger within the protagonist as she tries to figure out what the heck to do about the problem she's facing. That invisible, internal struggle is the third rail we've been talking about—it not only connects the novel's surface events to the protagonist's internal progress, giving those events meaning, but it's also what ultimately lets *you* know what those surface events will be (read: the plot).

This is why all stories are, by definition, character driven, and why the main character must come first. It doesn't matter what genre you're writing in—literary, historical, or speculative fiction, potboiler, or somewhere in between—the protagonist is the wellspring from which the plot will eventually flow. So before you can begin to figure out what that story is—let alone the plot—it stands to reason that first, you really need to know who your protagonist is.

Why Writing About Someone Isn't the Same as Writing About Anyone

I wouldn't be surprised if at this moment some of you are thinking, *Hello, I already told you who my protagonist is. He was right there in my What If, just like the unfortunate Romeo and Juliet were right there in Shakespeare's.* But the problem is, when we think of *Romeo and Juliet*, it's impossible not to envision the fully fleshed out characters we've all sobbed over (just *once* can't Juliet wake the hell up before Romeo gulps down that poison?). This phenomenon has been dubbed the Curse of Knowledge, which basically means that once you know something, it's well nigh impossible to imagine what it's like not to know it. But when it comes to an original protagonist, the most a *What If* gives us is a general description of someone, who could turn out to be almost anyone. The only things *R&J*'s star-crossed lovers came into the game with were a very strong mutual attraction and overly controlling parents (at least in Juliet's case). As characters they were completely unformed. Juliet could have been a mean girl and Romeo a cad; Juliet could have been a chambermaid and Romeo a page; and had Shakespeare knocked back a couple of cups of mead, Juliet could have been a squirrel and Romeo a penguin. The same is true of your protagonist. You're not looking for a general anyone, you're looking for a specific someone. A someone whose past will make what happens to them the moment they step onto the first page of your novel, inevitable.

- ❑ Gave subordinates the necessary resources to complete assignments most of the time
- ❑ Delegated fairly between subordinates
- ❑ Checked on assignment progress routinely most of the time
- ❑ Trained subordinates before making assignments ____% of the time
- ❑ Balanced assignments for subordinates between challenging and routine

❷ Below Standards (BS)

- ❑ Needed to perform better oversight of assigned tasks
- ❑ Showed reluctance to assign challenging tasks
- ❑ Intervened and meddled after assignments were made
- ❑ Failed to give clear instructions and support for assignments
- ❑ Interfered with progress of assignments by oversupervising
- ❑ Made abrupt and frequent changes to assignments due to lack of careful planning
- ❑ Failed to check on assigned work ____% of the time
- ❑ Kept the challenging assignments for himself most of the time
- ❑ Did not offer enough training or background information before making assignments
- ❑ Jumped back into projects after making assignments, thus undermining assignees' authority
- ❑ Did not assign projects in a timely manner
- ❑ Failed to give clear directions or goals in making assignments

❶ Far Below Standards (FBS)

- ❑ Failed to delegate as described in his job description
- ❑ Missed deadlines ____% of the time due to failure to delegate
- ❑ Stockpiled all complex projects for himself
- ❑ Demonstrated work quality that suffered due to failure to delegate
- ❑ Delegated only when he had not met deadlines, setting assignees up for failure

- ❑ Made assignments with very few or no instructions or support
- ❑ Misassigned projects to subordinates without skills to complete them
- ❑ Refused to delegate even routine tasks
- ❑ Lacked proper knowledge of subordinates' skill levels
- ❑ Did not supply the necessary resources and authority to subordinates given assignments
- ❑ Assigned projects inappropriately
- ❑ Failed to meet goals because of lack of delegation
- ❑ Had not trained employees to take on any of the department's new assignments

⓪ Not Observed (NO), Not Ratable (NR), Not Applicable (NA), Failed (F)

- ❑ Not applicable
- ❑ Demonstrates no aptitude for this role
- ❑ Failed to delegate
- ❑ Not observed

Recommended Action Steps for the Manager

- ■ Review the suggestions at the end of this section (see page 171), and adapt and use those that apply.
- ■ Ask the employee to analyze subordinates' skills and aptitude for future assignments.
- ■ Instruct the employee to query subordinates about their project interests.
- ■ Instruct the employee to increase the number of projects delegated in the next three months by ____%.
- ■ Have the employee list upcoming projects that may be delegated.
- ■ Assign the employee to prepare subordinates for projects through training.
- ■ Have the employee write up a plan for how she will effectively delegate.

- Ask the employee if he is willing to commit to being mentored by an employee with excellent delegating skills with the goal of increasing effective delegating by ____% in the next quarter.
- Recommend that the employee transfer to another job.

Dependability

Great job performance and dependability are inseparable when describing a valuable employee. This quality is essential to the efficient operation of every organization.

5 Outstanding (O)

- ❏ Could always be depended on to deliver as promised
- ❏ Took complete responsibility for own work
- ❏ Never let coworkers down
- ❏ Took total responsibility for department's quality production
- ❏ Delivered exactly as promised
- ❏ Completed all job requirements without supervision or reminders
- ❏ Possessed a perfect employee record
- ❏ Performed assignments on time and with ____% accuracy
- ❏ Supported coworkers and worked behind the scene to make sure quality was maintained
- ❏ Worked diligently to keep coworker morale high

4 Exceeded Standards (ES)

- ❏ Delivered as promised ____% of the time
- ❏ Took responsibility for own work
- ❏ Worked hard to not let coworkers down
- ❏ Completed most job requirements without supervision or reminders
- ❏ Showed nearly perfect employee records for punctuality and attendance
- ❏ Achieved a ____% accuracy rate in work assignments

- ❏ Had a good record for boosting coworkers morale
- ❏ Pitched in to help coworkers finish tasks to meet deadlines

❸ Met Standards (MS)

- ❏ Met acceptable standard of accountability for work
- ❏ Delivered on deadline ____% of the time
- ❏ Contributed consistently to department's total effort
- ❏ Achieved a ____% accuracy rate in work assignments
- ❏ Helped to build department morale
- ❏ Pitched in to help coworkers finish tasks to meet deadlines ____% of the time

❷ Below Standards (BS)

- ❏ Registered below acceptable level on delivering his work error free by ____%
- ❏ Delivered work late ____% of the time
- ❏ Needed the help of others to achieve a satisfactory accuracy rate of ____%
- ❏ Underachieved in attendance and punctuality by ____%
- ❏ Was a drain on the department morale

❶ Far Below Standards (FBS)

- ❏ Excused himself on issues of accountability for his work ____% of the time
- ❏ Racked up ____ corrective action marks for deficiencies in punctuality, attendance, department meeting attendance, and employee conduct
- ❏ Did not meet deadlines ____% of the time
- ❏ Failed to appear for special assignments ____ out of ____ times
- ❏ Did not meet ____ of ____ goals
- ❏ Needed constant assistance from supervisors to complete routine work

❶ Not Observed (NO), Not Ratable (NR), Not Applicable (NA), Failed (F)

❑ Not observed

❑ Failed to meet position dependability requirements

Recommended Action Steps for the Manager

■ Review the suggestions at the end of this section (see page 171), and adapt any that apply.

■ Discuss exact issues of deficiency to find out why the employee is missing the mark and address these.

■ Use solid goals for achievement to meet during the next quarter, have the employee list them explicitly, agree to meet them, and sign a personal contract.

Development of Subordinates

The employee who is tasked with developing his subordinates and does it well is very valuable to the organization. Weight this to reflect the portion of the employee's job it represents.

❺ Outstanding (O)

❑ Was dedicated to staff development and actively looked for new methods of increasing staff growth

❑ Promoted _____ staff members after conducting training programs

❑ Found training programs and helped inspire subordinates to enroll and excel

❑ Inspired and motivated subordinates to participate in training

❑ Conducted _____ programs of on-the-job training to develop subordinates' new skills

❑ Was an excellent role model for subordinates

❑ Recognized motivated subordinates and accelerated their training

❹ Exceeded Standards (ES)

- ❑ Promoted career development and growth among subordinates
- ❑ Promoted _____ of _____ subordinates in the past year
- ❑ Rated with a score of _____ by subordinates on the development scale
- ❑ Was approachable but still maintained supervisor relationship with subordinates
- ❑ Supported and reinforced new employee programs
- ❑ Made assignments to aid development among subordinates

❸ Met Standards (MS)

- ❑ Gave subordinates new assignments after they completed training
- ❑ Made subordinates aware of in-service training available
- ❑ Scheduled appropriate in-house training regularly
- ❑ Promoted _____ of _____ subordinates in the past year
- ❑ Was rated with a score of _____ by subordinates on the development scale

❷ Below Standards (BS)

- ❑ Did not institute training in a timely manner to help employees develop as quickly as possible
- ❑ Took a piecemeal approach to training, and coverage was not thorough
- ❑ Delivered inadequate training for new subordinates, resulting in their inability to perform to quality standards
- ❑ Did not allow for new techniques training
- ❑ Produced department errors _____% over acceptable rate due to lack of subordinate training

❶ Far Below Standards (FBS)

- ❑ Did not properly train subordinates
- ❑ Frustrated motivated subordinates by failing to give them challenging assignments
- ❑ Left subordinates on their own without adequate training

❏ Failed to promote any subordinate during the review period

❏ Did not conduct orientation for new subordinates

ⓞ Not Observed (NO), Not Ratable (NR), Not Applicable (NA), Failed (F)

❏ Had no development responsibilities

❏ Not Ratable

❏ Not observed

❏ Received a poor rating by subordinates for training

❏ Failed to develop subordinates

Recommended Action Steps for the Manager

- Survey employee's subordinates with him to learn how, specifically, training is failing.

- Insist that the employee ask subordinates for ideas on training they would like to have.

- Have the employee study work output records to learn how and where training needs to be improved.

- Review with the employee the training programs and make changes and upgrades.

- Have the employee write up a schedule that includes ____ training sessions every ____.

- Ask the employee to commit to implementing ____ changes to produce a score of ____ on subordinate evaluations by the end of the next quarter.

- Have the employee create a listing of outside sources of training and development to help motivated subordinates excel.

- Review the suggestions at the end of this section (see page 171), and adapt any that apply.

Equal Opportunity and Diversity

All employees must be accepting of their coworkers to ensure everyone has an equal opportunity to perform at their best. Employees who embrace these practices enrich the fabric of the workplace, raise morale, and avoid possible legal hassles.

❺ Outstanding (O)

- ❑ Excelled at offering equal opportunities to all subordinates
- ❑ Worked hard to ensure that subordinate population reflects that of diverse customer base and whole labor pool and population
- ❑ Supported the organization's diversity mission and outreach initiatives
- ❑ Encouraged contributions from all subordinates
- ❑ Conducted meetings to invite diverse ideas and points of view
- ❑ Treated all coworkers equally
- ❑ Built an environment of openness and respect for diverse ideas and opinions
- ❑ Was accepting and sensitive to all coworkers and subordinates
- ❑ Created a task force to promote EEO (Equal Employment Opportunity) practices

❹ Exceeded Standards (ES)

- ❑ Worked to conduct subordinate interaction without bias
- ❑ Promoted on the basis of demonstrated skill and achievement only
- ❑ Was diligent in maintaining equal pay for equal work among subordinates
- ❑ Embraced diversity as a strategic business initiative
- ❑ Established an open-door policy to hear all subordinate views and complaints
- ❑ Tried to identify labor pools that could supply diverse employee candidates

❑ Treated everyone in the work area equally

❑ Did not tolerate discrimination among subordinates

❸ Met Standards (MS)

❑ Used some diversity-based practices in seeking new employees

❑ Instituted a no-tolerance of discrimination policy in her work area with the help of HR

❑ Based promotion of subordinates on performance

❑ Hired new employees from diverse minority groups with the help of HR

❑ Did not demonstrate bias in interactions with subordinates

❑ Worked well with all coworkers

❑ Supported the values of diversity and EEO (Equal Employment Opportunity)

❷ Below Standards (BS)

❑ Needed to institute new practices of EEO (Equal Employment Opportunity) and diversity within department

❑ Did not have a proportionate representation of minorities among subordinates

❑ Had only white males in supervisory roles in department

❑ Refused to work with women and coworkers from minority groups

❑ Did not institute the organization's new program of creating a task force on diversity

❑ Showed some inequity in pay rates for subordinates

❑ Had insufficient number of minority workers in training programs

❶ Far Below Standards (FBS)

❑ Had ____ complaints of discrimination from subordinates

❑ Did not interview any minority applicants for openings

❑ Had ____ reports from subordinates that he made insensitive comments

❑ Told off-color jokes in department meetings

- ❑ Promoted and/or supported discrimination among coworkers
- ❑ Owned a record of inequitable pay rates among subordinates

ⓞ Not Observed (NO), Not Ratable (NR), Not Applicable (NA), Failed (F)

- ❑ Received _____ complaints of sexist, racist, and/or prejudicial comments
- ❑ Not ratable
- ❑ Refused to work with minority coworkers
- ❑ Failed to promote minority workers who were the most qualified for the positions
- ❑ Not observed

Recommended Action Steps for the Manager

- Ask the employee to hire women and/or members of defined minority groups for _____ out of the next _____ new hires.
- Have the employee write an EEO (Equal Employment Opportunity) and diversity plan for the department and list incremental steps and dates by which action will be taken.
- Insist that the employee apologize to subordinates for any inappropriate comments in the past, and announce a new policy of zero tolerance of discrimination and unacceptable comments and behavior going forward.
- Enroll the employee in sensitivity training to change improper attitudes and behavior.
- Have the employee meet with HR and study legal requirements for meeting EEO (Equal Employment Opportunity) and diversity guidelines and write a plan for complying.
- Ask the employee to establish a plan to recognize the contributions of all subordinates, and correct any inequities in remuneration.
- Have the employee commit to a plan of equal training for all subordinates and write the plan with dates for achievement.

Ethics

In today's diverse work environment, it can't be assumed that all employees share the same values and ethical standards. Therefore it's best to be proactive and include organizational rules and practices of ethical behavior on the job as part of the new employee training program. Rate the employee's performance carefully.

5 Outstanding (O)

❑ Maintained a perfect record and reputation for being scrupulously honest

❑ Emphasized ethical conduct in training programs

❑ Put ethical considerations above all considerations, including possible personal consequences, and presented a sterling role model

❑ Ensured that he adhered to the strictest code of conduct concerning conflict of interest

❑ Evaluated by customers and suppliers as holding the highest standard of integrity

❑ Exercised the highest principles of honest and open communications with subordinates

❑ Held his department to a high ethical standard

❑ Gained new customers and retained old ones through attention to treating them with honesty and integrity

4 Exceeded Standards (ES)

❑ Stressed strict adherence to the organization's code of conduct

❑ Stood for doing the right thing regardless of the consequences

❑ Held a reputation for fair dealing among customers

❑ Emphasized never bending the rules

❑ Rated very good in standing up for subordinates' rights

❑ Exhibited good adherence to ethics training with subordinates

❑ Studied and followed applicable laws and regulations

❸ Met Standards (MS)

- ❑ Held a good record of dealing fairly with customers
- ❑ Encouraged code of ethics among subordinates
- ❑ Did not try to bend the rules, find loopholes, or skirt ethical standards
- ❑ Knew the legal regulations for ethical behavior and followed them
- ❑ Was rated as "fair" by ____% of subordinates
- ❑ Received a "good" rating on fairness from ____% of customers

❷ Below Standards (BS)

- ❑ Bent the rules of ethics to improve production numbers
- ❑ Viewed ethics rules as too strenuous sometimes
- ❑ Found that strict adherence to the law was too confining at times
- ❑ Received "unsatisfactory" rating from ____% of customers on fairness issues
- ❑ Encouraged subordinates to cut ethical corners to improve profits
- ❑ Employed a double standard on matters of ethics for subordinates and himself
- ❑ Rated "unfair" in employee dealings by ____% of subordinates
- ❑ Ignored legal regulations on ____ occasions

❶ Far Below Standards (FBS)

- ❑ Failed to uphold standards of ethics
- ❑ Dealt unfairly with customers ____% of the time
- ❑ Did not behave ethically with customers
- ❑ Had ____ ethics violations
- ❑ Was responsible for legal action brought against the organization
- ❑ Infracted the organization's code of ethics

❶ Not Observed (NO), Not Ratable (NR), Not Applicable (NA), Failed (F)

- ❏ Not ratable
- ❏ Not observed
- ❏ Acted unethically in dealing with subordinates
- ❏ Did not admit a conflict of interest and tried to hide the results
- ❏ Lost ____ customers due to unethical behavior
- ❏ Lied to ____ and cost the organization ____ in lost customers and reputation

Recommended Action Steps for the Manager

- Encourage employee to admit the wrong behavior to those affected, apologize, and announce a commitment to change.
- Ask the employee to make atonement for wrongdoing where necessary, and write a commitment to change and a plan for the future—with precise actions and deadlines—and sign it.
- Insist the employee make a commitment to treat everyone with respect and provide everyone an with an equal opportunity.
- Ask subordinates for input on correcting policies that limit fair treatment and equal opportunities.
- Emphasize that the employee must avoid any perception of favoritism or preferential treatment of subordinates.
- Have the employee commit to begin a program to get to know subordinates and their needs in order to help them excel in their careers and advance within the organization.
- Be sure the employee holds all subordinates to the same high standard of ethical behavior.
- Require that the employee write up a plan for making the necessary changes for the next quarter, schedule a meeting within the next week to review and approve the plan, and have the employee sign the approved plan.

Flexibility

Being able to change directions quickly, seamlessly, and without rancor or major disruption is a valuable skill. Weight it according to the need for it within the employee's job description.

❺ Outstanding (O)

- ❑ Was always willing to work overtime to get a job done when the need occurred
- ❑ Could be counted on to accept and support necessary changes in policy and procedures; cooperate, and compromise
- ❑ Anticipated situations that called for quick procedural changes and incorporated them promptly
- ❑ Was versatile and able to manage multiple projects at the same time, shifting gears quickly and seamlessly
- ❑ Changed production procedure within ____ hours to eliminate errors
- ❑ Dealt easily with interruptions and navigated obstacles in a professional manner
- ❑ Was willing to consider alternatives that might improve outcomes
- ❑ Handled emergencies effectively

❹ Exceeded Standards (ES)

- ❑ Stepped in quickly and easily to fill a gap when coworkers needed help
- ❑ Employed an array of alternate techniques and procedures to keep work on schedule
- ❑ Shifted focus and approach easily
- ❑ Filled in for absences whenever necessary
- ❑ Was able to change priorities when directed by manager

❸ Met Standards (MS)

- ❑ Cooperated, compromised, and usually made the necessary changes when directed
- ❑ Reset priorities when required

- Was able to shift gears and step into a number of tasks with less than ____% loss of production
- Adjusted to changes in procedures with a minimum of downtime
- Changed schedules and assignments to meet deadlines
- Accepted required changes in work assignments from superiors with minimal resistance

❷ Below Standards (BS)

- Resisted necessary changes in work assignments ____% of the time
- Needed convincing when directed to make changes
- Had a record of ____% loss in production when he had to implement procedure change
- Struggled with resetting priorities, cooperating, and compromising
- Fought any change, even when evidence showed it was the best course
- Delayed implementing changes when given a deadline

❶ Far Below Standards (FBS)

- Demonstrated inflexibility when presented with any change
- Refused to change until threatened with disciplinary action
- Was unable to change priorities ____% of the time
- Was listed inflexible by ____ of ____ subordinates
- Showed an uncooperative attitude with coworkers in instituting changes
- Exhibited inflexibility that lost the organization ____ customers

⓿ Not Observed (NO), Not Ratable (NR), Not Applicable (NA), Failed (F)

- Not observed
- Failed to demonstrate the necessary flexibility for the position
- Not ratable
- Not applicable

- Review the suggestions at the end of this section (see page 171), and adapt them to help the employee with flexibility.
- Have the employee write a plan for himself that outlines specific areas where he can improve his flexibility, outline the steps he will take to accomplish this, and then commit to making those changes in the next quarter.
- Ask the employee to agree to taking on new areas of responsibility or tasks and sign a personal contract to do so.

Forward Thinking

The employee who is able to project ahead and anticipate the needs, attitudes, and reactions of those with whom he must interact can perform at a higher efficiency rate by eliminating the need to repeat the decision-making steps, coordination steps, and reworking of projects. For some positions, this skill should be rated quite high.

⑤ Outstanding (O)

❑ Anticipated with nearly 100% accuracy how decisions would be received by coworkers, customers, and the public

❑ Held a ____% success-rate record for anticipating possible problems with procedural changes and developed contingency plans in advance

❑ Followed industry trends and contributed ideas that pushed the envelope

❑ Used futuristic thinking and planning to help the organization stay ahead of the curve

❑ Projected accurately the possible consequences of decisions

4 Exceeded Standards (ES)

❑ Anticipated with ____% accuracy how decisions would be received by coworkers, customers, and the public

❑ Held a ____% success-rate record for anticipating possible problems with procedural changes and developed contingency plans in advance

❑ Used the input of others and brainstorming sessions designed to help with anticipating possible problems and developing remedies before new procedures were put in place

❑ Used industry trends to inform her decision making

❑ Tried to stay ahead of the curve by forming task forces to plan for new products and procedures

3 Met Standards (MS)

❑ Was helpful to superiors in predicting how changes would be received by coworkers

❑ Contributed some valuable input on task forces charged with brainstorming new products and procedures

❑ Anticipated with ____% accuracy possible problems with procedural changes and helped to develop contingency plans in advance

❑ Responded well when changes needed to be anticipated and procedures planned to minimize disruptions in workflow

2 Below Standards (BS)

❑ Was unable to anticipate consequences of procedural and policy changes

❑ Needed assistance to institute changes after new procedures resulted in unexpected consequences

❑ Let change happen without proper planning and struggled with the aftermath

❑ Made a minimal contribution to task-force brainstorming

❶ Far Below Standards (FBS)

- ❑ Was surprised and unprepared for consequences of procedural changes
- ❑ Did not contribute to task-force brainstorming activities
- ❑ Failed to keep coworkers and other departments informed about new procedures
- ❑ Had ____ work stoppages due to unanticipated problems after new procedures were put in place
- ❑ Was unable to anticipate problems or institute changes

❶ Not Observed (NO), Not Ratable (NR), Not Applicable (NA), Failed (F)

- ❑ Not applicable
- ❑ Not ratable
- ❑ Not observed
- ❑ Failed to make any contribution to anticipating repercussions of new procedures

---

Recommended Action Steps for the Manager

- ■ Review the suggestions at the end of the section (see page 171) and adapt any that will help the employee increase her skills in this area.
- ■ Establish a program for rewarding the employee who has exemplary skills of anticipation and forward thinking.
- ■ Make sure the employee understands how important this skill is to his chances for promotion and career development.

Goal Setting

An employee needs concrete and tangible career goals that are properly focused on the organization's goals. If he lacks these goals, he is like the driver who has neither a GPS nor a road map. It is the manager's job to help his subordinates remain properly focused on the organization's goals, and then to help each employee set realistic and attainable goals for his contribution. And it is also the manager's role to help his subordinates achieve those goals. Weight this skill according to the demands for the employee's contribution in this area.

❺ Outstanding (O)

- ❑ Was focused and persistent in pursuing organizational and individual goals
- ❑ Was proactive in setting achievable, specific, measurable, relevant, and time-specific goals for himself
- ❑ Inspired subordinates to set and achieve relevant personal goals
- ❑ Maintained goal focus throughout diverse steps and procedures ____% of the time
- ❑ Understood the importance of goal setting in the achievement process, and had a ____% success rate in reaching goals for the past ____
- ❑ Held subordinates to a high standard of meeting their own set and measurable goals
- ❑ Implemented a system of goal setting that included objective methods of measuring and rewarding success
- ❑ Communicated organizational goals to subordinates and translated these into measurable achievements for his department
- ❑ Created performance standards for subordinates
- ❑ Was ____% successful in getting subordinates to reach department goals

❹ Exceeded Standards (ES)

- ❑ Was realistic in setting achievable goals for department
- ❑ Communicated organizational goals to subordinates successfully ____% of the time
- ❑ Implemented performance standards for subordinates

- Was ____% successful in getting subordinates to sign on to goals programs with concrete performance standards
- Convinced ____% of subordinates to set and achieve their own goals in a ____ -month period
- Translated organizational goals into departmental tasks with ____% success
- Did not lose goal focus
- Inspired his subordinates to reach their goals

❸ Met Standards (MS)

- Complied with goals set by superiors ____% of the time
- Adjusted goals as required by superiors
- Implemented changes required to meet organizational goals ____% of the time
- Set realistic goals for his subordinates
- Worked hard to meet assigned goals and was ____% successful
- Changed goals when necessary
- Was successful in setting short- and long-term goals for department

❷ Below Standards (BS)

- Needed to take a more active role in setting goals for department
- Was unable to set goals with the necessary organizational focus without assistance
- Relied on supervisor to communicate goals
- Lacked self direction and initiative in setting goals for herself
- Resisted supervisor's attempts to set goals
- Had trouble translating organizational goals into work procedures
- Failed to break down long-term goals into short-term steps
- Set goals without consideration to specificity and measurability

❶ Far Below Standards (FBS)

☐ Demonstrated a disconnect between organizational goals and his work

☐ Resisted all attempts to set meaningful goals

☐ Could not translate long-term goals into everyday work procedures and deadlines

☐ Failed to meet the goals set in conjunction with his supervisor, ____% of the time

☐ Did not accept goals set by his supervisor

⓿ Not Observed (NO), Not Ratable (NR), Not Applicable (NA), Failed (F)

☐ Failed to set long- or short-term goals as requested

☐ Not observed

☐ Not applicable

☐ Not ratable

☐ Reached only ____ of his ____ personal goals

Recommended Action Steps for the Manager

- Review the suggestions at the end of the section (see page 171) and adapt those that are applicable.

- Review organizational long- and short-term goals with the employee and ask her to write personal goals to implement in her job description.

- Require the employee to write a set of goals for himself and his department and/or subordinates for the next quarter, with specific achievements and dates, and sign it.

Industry and Quality Work Habits

Measuring work output or industriousness and demonstrating sterling work habits show that not only can the employee make an exemplary contribution to the achievement of the organization's goals, but he can also set the standard for other employees. This "soft skill" exhibited by one employee can raise the bar for others and inspire them to reach a new level of accomplishment and performance. Weight this skill to reflect the importance of the employee's contribution.

❺ Outstanding (O)

- ❑ Saw every task through to completion
- ❑ Had an internal gauge that did not allow her to do less than her best
- ❑ Took no task shortcuts when it came to quality of work
- ❑ Approached each task with enthusiasm
- ❑ Did not leave work undone
- ❑ Finished every task well before the deadline
- ❑ Assisted others to ensure every part of the job was done to a high standard
- ❑ Insisted that each task be accomplished to the highest standard
- ❑ Paid very close attention to quality and elimination of errors
- ❑ Made sure coordination with coworkers was complete

❹ Exceeded Standards (ES)

- ❑ Saw most tasks through to completion
- ❑ Produced work that met a high quality standard
- ❑ Took shortcuts only when pushed to do so
- ❑ Showed enthusiasm for his job most of the time
- ❑ Went beyond job requirements to help others complete their tasks
- ❑ Did not procrastinate or delay taking on work at hand
- ❑ Had good work ethic
- ❑ Set a good example for coworkers in accomplishing tasks

❸ Met Standards (MS)

- ❑ Completed projects on time ____% of the time
- ❑ Saw tasks through to completion without supervision ____% of the time

- ❑ Met quality control standards ____% of the time
- ❑ Took shortcuts to try to meet deadlines
- ❑ Contributed to a good work environment
- ❑ Helped others complete their tasks occasionally
- ❑ Did not join in on time-wasting activities

❷ Below Standards (BS)

- ❑ Did not complete projects on time ____% of the time
- ❑ Did not meet quality of work standard ____% of the time
- ❑ Worked consistently on projects only with reminders
- ❑ Took shortcuts frequently (____% of the time) to try to meet deadlines
- ❑ Had ____ complaints for bad attitude and conflicts with coworkers
- ❑ Refused to help others in order to meet own work production deadlines
- ❑ Was written up for time-wasting activities on the job ____ times

❶ Far Below Standards (FBS)

- ❑ Did not complete projects on time ____% of the time
- ❑ Failed to meet the minimum quality control standards ____% of the time
- ❑ Needed to be closely supervised to produce the minimum output
- ❑ Could not work independently
- ❑ Delayed starting work until deadline was looming, and then took wasteful shortcuts
- ❑ Evaluated by coworkers as a troublemaker in ____ out of ____ reports
- ❑ Was cited for wasting time ____ times

❶ Not Observed (NO), Not Ratable (NR), Not Applicable (NA), Failed (F)

- ❑ Failed to meet minimum standard
- ❑ Not observed
- ❑ Not ratable
- ❑ Not applicable
- ❑ Must be terminated due to _____

- Review the suggestions at the end of the section (see page 171) and adapt any that will help the employee improve.
- Have the employee write his own plan outlining how he can increase his value to the organization and the team through personal changes in his work habits.
- Institute a bonus program that rewards industry and good work habits, being sure to spell out precisely what these habits are in terms of the work environment.
- Be sure the employee understands how critical excellent work habits are to being promoted; have her write a plan to improve within this quarter, and sign it.
- Use an ongoing program of recognizing outstanding work habits on a monthly basis.

Initiative

Some employees seem to have almost a sixth sense about what needs to be done and when. They are the people who can look at a task or course of action and anticipate what is best done next, and then take the initiative action to make sure it happens effectively. Weight this skill generously when it results in the employee doing an outstanding job.

5 Outstanding (O)

- ❏ Looked for opportunities to take the reins
- ❏ Took the proper action at the right time every time
- ❏ Negotiated a ____% discount with suppliers, saving the organization $_____ per year
- ❏ Recalibrated the production equipment to reduce waste by ____%
- ❏ Foresaw and averted a production slowdown by locating new suppliers

- ❏ Reworked the flow of reports to enhance communications by ____%
- ❏ Wrote new policy and procedures for office online communications and avoided organization leaks
- ❏ Anticipated problems and proactively solved them before they happened
- ❏ Was self-motivated in setting and meeting personal goals
- ❏ Helped out in areas not his responsibility
- ❏ Communicated effectively to ensure projects were done correctly and on time
- ❏ Took proactive steps to make sure his team met organization objectives
- ❏ Averted a client crisis by effectively communicating delivery schedule changes

❹ Exceeded Standards (ES)

- ❏ Took initiative ____% of the time
- ❏ Averted problems frequently by taking preventive steps
- ❏ Took proactive steps frequently to streamline production
- ❏ Demonstrated good skills of anticipation and initiative
- ❏ Reworked the flow of reports to enhance communications by ____%
- ❏ Negotiated a ____% discount with ____ suppliers
- ❏ Came up with policy change suggestions to avoid employee conflicts
- ❏ Strived to improve the work process

❸ Met Standards (MS)

- ❏ Took initiative ____% of the time
- ❏ Averted problems by taking preventive steps ____% of the time
- ❏ Took proactive steps often to streamline production
- ❏ Demonstrated skills in anticipating needed changes in procedure ____% of the time
- ❏ Handled problem-solving decisions appropriately
- ❏ Preferred acting within written job description, but sometimes took added initiative

- Showed some risk-taking ability to solve customer problems
- Took necessary steps to create a backup system for production
- Initiated interdepartmental communications to avoid duplication
- Coordinated with other departments at an acceptable level

❷ Below Standards (BS)
- Showed discomfort in taking initiative
- Did only prescribed job description
- Was slow to act when faced with operational decisions
- Initiated nothing outside standard procedures
- Made decisions only with approval
- Did not foresee or take action to avoid problems in meeting delivery schedule ____ times
- Lacked confidence to initiate new procedures
- Failed to step up this part of his game to initiate needed changes

❶ Far Below Standards (FBS)
- Functioned far below job demands for taking initiative
- Failed to take initiative when a crisis called for action
- Did not take initiative until required to do so
- Delayed making the decision to act until it was too late
- Missed opportunities to avert customer problems because of lack of initiative
- Waited for procedure failure before acting
- Called often on the manager for a decision before taking action

⓿ Not Observed (NO), Not Ratable (NR), Not Applicable (NA), Failed (F)
- Not ratable
- Not observed
- Needed to be transferred to another department where this skill was not required
- Was unable to deliver the initiative required

- ❑ Was unable to function at the job description level
- ❑ Consistently failed to take initiative

Recommended Action Steps for the Manager

- Review the suggestions at the end of the section (see page 171) and adapt any that will help the employee improve.
- Be sure the employee understands how valuable taking the initiative is in order for her to be promoted.
- Have the employee write his own plan outlining how he will increase his value to the organization and his team by taking the initiative in key situations, explaining exactly when, where, and how he will do this within this quarter, and sign it.

Innovation

Creating a new way to get to a goal is a quality that can be extremely valuable, particularly in key positions where methods and measures have topped out in their efficiency. Thinking outside the box in coming up with new solutions to problems and devising better ways of doing things can help the organization reach new heights in productivity and efficiency. Weight this skill based on the employee's need for its application.

5 Outstanding (O)
- ❑ Could be depended upon to head up new methodology and approaches to problems
- ❑ Came up with efficiencies and streamlined procedures without prompting
- ❑ Was always thinking about better ways to do a task
- ❑ Reduced production costs by ____% by eliminating ____ unnecessary steps in the process
- ❑ Increased production by ____% by making ____ production line changes

- ❏ Combined ____ departments into ____, which resulted in better communication and subordinates indicating a ____% increase in job satisfaction
- ❏ Instituted a new bonus plan, which has resulted in an increase of ____% in subordinates' production and a ____% decrease in errors

❹ Exceeded Standards (ES)

- ❏ Brought innovative ideas to the manager regularly without prompting
- ❏ Developed a better system to handle customer complaints and follow-up, which resulted in a ____% increase in customer satisfaction
- ❏ Was often thinking about improved ways to reach the department's goals
- ❏ Instituted a bonus program for subordinates to reward them for innovation that resulted in better production and higher customer and employee satisfaction
- ❏ Volunteered to head up searches for new methodologies and approaches to problems

❸ Met Standards (MS)

- ❏ Responded when asked to develop a better procedure
- ❏ Came up with improved ways to reach the department's goals ____ times
- ❏ Developed a new bonus program when assigned the task
- ❏ Did an acceptable job heading up a committee to search for new methods and approaches to a production problem
- ❏ Served on the interdepartmental committee for procedure review and action

❷ Below Standards (BS)

- ❏ Resisted considering new ways to streamline present processes
- ❏ Responded to directives to come up with new problem-solving methods only when given a deadline

- ❏ Failed to produce new efficiencies that were suggested and demonstrated by coworkers
- ❏ Failed to accept and implement new innovations when they were presented

❶ Far Below Standards (FBS)

- ❏ Refused to consider new ways to streamline present processes in more efficient ways
- ❏ Did not develop any problem-solving methods, even when asked to do so
- ❏ Stuck to old methodologies, even when new innovations were proven to increase production, employee morale, and reduce waste
- ❏ Stonewalled attempts to introduce new innovations

❶ Not Observed (NO), Not Ratable (NR), Not Applicable (NA), Failed (F)

- ❏ No applicable
- ❏ Not ratable
- ❏ Does not possess the necessary innovation skills to function in this position
- ❏ Failed to innovate as required
- ❏ Needed to be transferred to another position

Recommended Action Steps for the Manager

- Review the suggestions at the end of the section (see page 171) and adapt any that apply.
- Create a mentorship for the employee with another employee who excels in this skill and require the employee show a ____% improvement in the next quarter.
- Have the employee write up a plan outlining how she will increase his innovation skills in exact terms with dates, and sign the agreement to do this.

Interpersonal Skills

The employee everyone wants to work with is the employee who helps create a work environment that is harmonious and productive. Weight this skill higher when there is the need for close working relationships and lots of interaction among employees.

❺ Outstanding (O)

- ❑ Exuded positive attitude and magnetism that opened up those with whom she interacted—customers, suppliers, coworkers, subordinates
- ❑ Selected first for team participation
- ❑ Exhibited the brand of leadership that got cooperation from others
- ❑ Received praise for attitude, contribution, and likeability
- ❑ Resolved conflicts between subordinates and coworkers easily and quickly
- ❑ Maintained excellent interdepartmental relationships
- ❑ Motivated others to do their best
- ❑ Accepted input and suggestions from everyone equally
- ❑ Practiced model diversity and EEO (Equal Employment Opportunity) principles
- ❑ Showed an authentic interest and care for those he worked with

❹ Exceeded Standards (ES)

- ❑ Held a reputation for getting along with everyone
- ❑ Selected first for team participation
- ❑ Showed leadership in many situations
- ❑ Demonstrated an ability to get others to open up
- ❑ Maintained good interdepartmental relationships
- ❑ Was effective in motivating others to produce at a higher level
- ❑ Practiced diversity and EEO (Equal Employment Opportunity) principles consistently
- ❑ Initiated help projects for ____ colleagues who suffered prolonged illness

❸ Met Standards (MS)

- ❑ Used proper language in the workplace—absent of sexual or racist content
- ❑ Had only ____ subordinate complaints registered in the last ____
- ❑ Could usually get subordinates to follow his lead
- ❑ Was effective in settling disputes ____% of the time
- ❑ Met expectations in communicating with subordinates and coworkers
- ❑ Was often one of the first selected as a team member

❷ Below Standards (BS)

- ❑ Had difficulty in getting the support of her subordinates
- ❑ Could only get a minimum number of subordinates to follow his lead
- ❑ Had ____ subordinate complaints registered against him in the last ____
- ❑ Needed to have assistance in settling disputes ____% of the time
- ❑ Was selected last to serve on teams
- ❑ Exhibited deficits in communication skills
- ❑ Was required to take sensitivity training for ____ complaints by subordinates but showed no improvement

❶ Far Below Standards (FBS)

- ❑ Required disciplinary action ____ times for improper language and sexual comments
- ❑ Had unacceptable number (____) of complaints by customers, clients, coworkers, and/or subordinates in the last ____
- ❑ Was not able to function at the minimum level to resolve disputes
- ❑ Contributed to disputes ____ times in the last ____
- ❑ Was unable to communicate properly with subordinates
- ❑ Was rejected by ____ coworkers as a team member

O Not Observed (NO), Not Ratable (NR), Not Applicable (NA), Failed (F)

❏ Failed to exhibit the necessary skills to function in this position

❏ Not observed

❏ Not ratable

❏ Needed to be transferred to another position

Recommended Action Steps for the Manager

- Review suggestions at the end of this section (see page 171) and adapt as many as practical to help the employee improve.

- Require the employee to complete sensitivity training classes.

- Present leadership and communication skills classes to employee and explain the need for her to take them.

- Have the employee write a plan for how he will improve in deficit areas over the next quarter, and sign it.

Job Knowledge

The employee who has a broad-based knowledge of all the factors that affect her job, in addition to the actual "hard skills" and knowledge needed to perform it, will be able to achieve a much higher level of accomplishment than she otherwise would; and she will be able to advance quickly. This is an important basis for excelling within any organization, and should be weighted and recognized according to the demonstrated benefit the organization has received.

❺ Outstanding (O)

❏ Possessed an in-depth knowledge of the critical issues, well beyond his job responsibilities, that affect the entire organization

❏ Had a rare and extensive knowledge base that enabled her to perform above his job description

❏ Understood the purposes, objectives, practices, and procedures of the entire department

Here's how our brilliant brain *creates* the world we see:

Day by day, from birth on, as we interact with our immediate surroundings, our hungry brain swills down useful info that it interprets as
the way of the world. But here's the game changer: what we're learning
isn't "objective," so that everyone comes away with the exact same interpretation of what things mean. Rather, it's all learned *subjectively*, based on
personal experience, so everyone has a *different* interpretation of the same
"objective" thing. Point being: Meaning is always subjective, so even when
on the surface we agree on what things mean, it's often for vastly different
reasons (or, just as likely, one of us is wary of saying what we really think).

We ascribe meaning to everything—home, clouds, love, and the fact
that our significant other forgot our birthday *again*—based on one thing
only: what our *personal experience* has taught us that those things signify,
and therefore what we can expect of them.

As cognitive scientist Benjamin K. Bergen points out in his revelatory book *Louder Than Words: The New Science of How the Mind Makes
Meaning*, each of us assigns different significance to things because we
all use our own subjective mental perceptions to construct meaning. He
writes: "We all have different experiences, expectations and interests, so we
paint *the meanings we create* for the language we hear in our own idiosyn-
cratic color" (italics mine).[1]

But what does that mean exactly? Let's try an exercise Bergen himself suggests. I'm about to give you two words, and when I do, I want you
to close your eyes and form a full picture of the first thing that comes to
mind—is it day? night? inside? outside?

Okay, here are the words: barking dog.

When I've done this exercise in writing workshops I've gotten answers
as varied as these:

- I saw the huge pit bull with the sharp yellow teeth who lunged at me
  when I was coming home from school in my gingham pinafore when
  I was six.

- I saw my old basset hound, Fred, who always barked his head off and did a clumsy little happy dance when I came home from work. Oh, how I miss that dog.
- I saw my neighbor's Jack Russell terrier, who barks all damn night. I haven't slept in weeks, I'm behind at work, I keep fighting with my wife . . . oops, did I say that out loud?

The point is, everyone sees a very different image. Some of us see a German shepherd with a loud terrifying bark, some a pampered Pomeranian on a silk pillow, and some our own loving, tail-wagging mutt whose bark of delight when we get home is what makes life worth living.

And notice that it's not just the image we see that's different; it's also how that image makes us feel, what we do in response, and how it's impacted our life, not to mention our belief system. If you were attacked by a dog when you were little, you may have religiously avoided dogs, and dog lovers, ever since. Or, if you saw good ol' Lassie running in slo-mo through an amber field of grain, right now you might be sniffling as you revise your will to leave everything to PETA, your greedy relatives be damned.

There is only one thing that no one *ever* sees upon hearing the words "barking dog":

"A highly variable domestic mammal (*Canis familiaris*) closely related to the gray wolf."

That, my friends, is the dictionary definition of a dog—aka an "objective" general fact. The point is, we never see or do *anything* "in general." It's actually kind of impossible, when you think about it. For instance, did you ever go to school in general? Go to the market in general? How about fall in love, in general? Of course not, because we do every single thing, specifically, moment by moment.

But at least when you hear "barking dog," you readily picture something that actually is a dog. With conceptual notions like "torture chamber" or "bed of roses," the difference between the general and the specific is even more pronounced—you can fill in the blank with just about anything, but at least you *can* fill it in with a concrete, if random, visual.

Completely abstract terms like love, loyalty, hate, or trust are even *more* fuzzy, because they're completely, totally conceptual. Each of us is going to have a different interpretation of what they mean, different images, different rules of engagement, different beliefs, and different conclusions. These concepts—which writers are often encouraged to offer up as their theme—are only general categories, placeholders. By themselves, they're a big, empty "yeah, and so, what's your point?" Because the story, as I'm very fond of saying, is in the specifics. And the specific always comes back to your protagonist. The question isn't what does loyalty mean in general, but what does it mean to *her*? What is she loyal to? Why? What does it cost her? To figure that out, first we have to answer the question: what does "specific" *really* mean?

How Specific Is Specific?

Remember how specific all those "whys" you nailed down in the last chapter seemed? You did the exercise exactly right, but by definition, those whys are still too general; until we dig deeper, they can't be anything else. That's fine, because they've fulfilled their purpose. They're clear enough to show you where to dig for the precise specifics that will fuel your story. But what, exactly, is a specific?

Specific refers to what happened—literally. Not a general summation or overview of it, but the event itself, moment by moment, as it unfolded. The problem is that it's very easy to think you've nailed the specific when what you've really created is a convincing facsimile—or what I like to call a "general specific." And as with cubic zirconia, fool's gold, and a Mark Landis version of a Picasso, it's embarrassingly easy to mistake it for the real thing. Especially now, when you're starting from scratch, and everything you've come up with about your protagonist feels specific—that is, as compared to knowing nothing at all.

For instance, I once worked with a writer whose main character, Wanda, was an ace assassin who was trying to get out of the business.

Sounds pretty specific, doesn't it?

"Wow," I said, "that's an interesting choice. What made her want to be a hit woman in the first place?" The writer looked at me, blinked, took a deep breath, and then paused for way too long. It was clear he'd never really thought about it. Finally he said, "Um, because she has abandonment issues?" Right. Show me the person who says they don't have abandonment issues, and I'll show you someone who hasn't figured it out yet. We all have abandonment issues.

The first question he needed to answer, then, is: why, specifically, did Wanda have abandonment issues? Because the event that triggered her fear of abandonment will tell us a lot about what *she* sees as abandonment, and how that ultimately led her to kill people for a living.

For instance, it might be that Wanda's parents cared about her deeply and showed it at every turn. Except this one time, when Wanda was in the seventh grade and she got the lead in the school play. Her dad promised he'd be there, but he got a flat tire on the way to school and was ten minutes late. So when Wanda stepped onto the stage and looked into the audience, the seat she'd reserved for him was empty. Her heart fell, and she's been unable to trust anyone since. Now, that tells us something about Wanda, and how little it takes to earn her distrust.

Or how about this: Wanda's been concerned that lately her parents have seemed worried and distracted. Walking home from school, she tells herself that when they see her report card, with straight As, everything will be okay. But when she turns onto her block, she discovers that her house has been razed, her parents are gone, and none of the neighbors can look her in the eye as Social Services takes her away. In that case, Wanda is a very different kind of person, with another kind of abandonment issue.

But here's something that might surprise you as much as it did Wanda's author. Those two scenarios are helpful, but they're *still* way too general. Yes, they each pinpoint the specific event that triggered Wanda's "abandonment issues." But they're about what happened, externally; neither gives us a clue how it affected Wanda, internally—we have no idea what her beliefs were up to that moment, what they were based on, or how they changed in the face of what happened. In other words: We know nothing about the conclusion she drew about human nature in that moment, which has been her default belief ever since. The goal isn't to show us *that* she's changing; the goal is to show us what, specifically, she's changing *from* and what she's changing *to*—internally.

Just as the protagonist's POV isn't like a camera lens, neither is a scene written as if you're narrating something that you're watching on a video screen. Instead, you want to plant us inside your protagonist's head as the event unfolds. Here's the secret: being able to see it through your protagonist's POV means letting us hear what she's thinking as it happens—and not what she's thinking in general, but her struggle to figure out what's going on and what the hell to do about it. These thoughts will be woven throughout every paragraph in your novel.

Digging to the Center of Your Story: The Origin Scene

You've already pinpointed your protagonist's misbelief, and chances are you have a general idea what caused it—some scene or event or trauma. Jennie, for example, is working on a scene in which someone close to her protagonist dies, leading her to believe that love isn't worth the pain. Since the wisp of an idea that originally caught her attention was about grief, and about the way love and death play off each other, this makes perfect sense.

Here's Jennie working it out:

> Okay, so I don't want it to be her mom or her dad because that's so Disney. Having it be a sister or a friend—I don't know. That feels different than what I want. I can't explain why. I'm thinking *boyfriend*? No, too on the nose. Okay, so then someone older, maybe, whom she likes and trusts—like a teacher. Wait, I know—a best friend's mom or dad. Whom she felt almost closer to than her own parents. Whom she envied her friend for. I like that. I remember that feeling—I had a friend whom I thought had the perfect family. I was kind of obsessed with them, especially when my parents were going through a bad divorce. And maybe

this dad dies right after promising Ruby something— like he's going to read what she wrote.

I know, I know—you're going to ask for a specific scene, right? Okay, so the dad dies, and it's very strange for Ruby because her friend—I'm going to call her Beth—and Beth's family is so rocked. Ruby is waiting for everything to go back to normal, and she thinks it's going to happen at a weekend soccer game—the girls are on the same team. But it's actually the most brutal moment of all—because the dad was Beth's biggest fan and *he's not there.* When Beth crumbles, Ruby sees exactly what all that love cost that perfect family once it's gone.

That's a solid beginning, and exactly the sort of specific moment you're looking for. Remember, this is not the scene itself, yet. It's the rough general sketch that you'll soon expand into a full-fledged scene. What gives Jennie's sketch power is that it's a moment that completely shatters Ruby's expectations. She longed for a loving family, just like her friend's. But when the dad dies, Ruby sees the very real hole his absence leaves in the family, one that can't easily be filled and that seems to change the family completely. And just as she'd never seen love like that before, neither had she seen the excruciating pain that's left in its wake. That's when she realizes that love comes at a very steep cost. Her takeaway: love is not worth the price. A scene like this will definitely change Ruby's worldview. Not for a day or two, but long into the future, as it becomes the lens through which she sees the world.

WHAT TO DO

Now it's your turn. Can you envision the moment in your protagonist's life when his misbelief took root? Perhaps you have a vague idea, and that's fine. Take a minute and sketch it out the way Jennie did in a simple paragraph. If you're writing a mystery, courtroom drama, or police procedural, your focus might be as much on the crime as on the investigator, if not more so, thus you might do this for the bad guy *and* for your sleuth, whether she is a detective, lawyer, investigator, spy, or curious next-door neighbor. Remember: every character filters the world through his or her own internal logic, based on what the events in their past forced them to face. The goal is to find the defining moment in their past.

Well, What Did You Expect? Yep, There's Always a Before

Your next goal will be to transform this life-altering turning point into a full-fledged scene—so you know not only what happened, but exactly how your protagonist made sense of it as it did. It might be a huge event, like Wanda discovering her house has been razed. Or something tiny and seemingly mundane, like Wanda's dad arriving at her student play ten minutes late. It's not the external dramatic scope of the moment that matters; it's your protagonist's internal reaction to it that counts. Your main job is to track how her viewpoint changes throughout the scene. She will go into the scene believing one thing, and come out believing something else altogether. You're going to show us how the new misbelief clobbers the old one.

Um, wait, you may be thinking, *doesn't that mean I need to know a bit about my protagonist before I can write this scene? Otherwise, how will I know exactly what that old belief even is?* You're absolutely right. Even if she's only six, she already has a worldview that, like most of us, she's never even questioned—why should she? As far as she can see, it's "just the way things are." Boy, is she in for a rude awakening. Point being, if she doesn't enter with an established worldview, how can what's about to happen change it?

So, yes, even as you create this defining moment, you'll still be reaching into the past and gathering a handful of specifics in order to set it up. That's why before you begin writing the scene, you need to answer four questions. These are the same questions you'll ask yourself when writing— or envisioning—*any* scene. They are

- What does my protagonist go into the scene believing?
- Why does she believe it?

- What is my protagonist's goal in the scene?
- What does my protagonist expect will happen in this scene?

Let's quickly go through them one by one.

What Does My Protagonist Go into the Scene Believing?

You've already identified the misbelief that your protagonist will struggle with throughout the novel. The question is: what existing belief did it topple in order to take root in the first place? Remember, your protagonist isn't going from "neutral" to a new belief. She enters already believing something, a belief she will struggle mightily to hold onto during the scene. In Ruby's case, that belief is that loving families are not only possible, they're what keep you safe.

Why Does She Believe It?

The question is, What, specifically, instilled your protagonist's *old* belief? Since it's probably something she never even thought about, you, as sleuth, must dive into her life and extract one moment that exemplifies this belief to her. The good news is that there are probably many moments you could choose from, but all you need is one. For instance, if her old belief is that you can always count on your parents to be there for you, it might be the time when she was sick, and her dad stayed home from work, made her chicken soup, and read to her all day. What you're looking for is the specific memory she'll instantly summon up when the misbelief smacks her in the face, and she thinks, wait, that can't possibly be true, because . . . .

What Is My Protagonist's Goal in the Scene?

Ask yourself, what does my protagonist go into this scene wanting to achieve by the time it ends? It helps to keep in mind that in every scene you ever write, the protagonist must enter with a goal. This isn't a writing convention; it's life. Without a goal, we'd slip into stasis, and hey, even stasis has a goal: to keep things exactly as they are, forever. Which, let's face it, takes work.

The question now is, What does your protagonist want the outcome of this scene to be, *not* in general, but specifically? This means nailing down more specifics about the scene.

What does Ruby want? She wants to find proof that loss, even great loss, doesn't have to change anything. That is to say, she wants the one thing that no one ever gets: for everything to go back to the way it was.

What Does My Protagonist Expect Will Happen in This Scene?

Regardless of what we humans *want* to happen, we go into every situation with a clear set of expectations of what we think *will* happen. How else could we gauge the meaning of what *does* happen? In a scene this is often where the real conflict and tension come in, as we watch the protagonist struggle to reconcile what she expected would happen with what's actually happening—while trying not to let anyone see her sweat. That's why *you* not only need to know what your protagonist expects going into the scene, but you have to be sure you've made it clear to your readers as well. If we don't know what the protagonist expects, how will we know if those expectations are met? Or, more likely, completely upended?

But here's something interesting: when I say "upended," I don't necessarily mean that, on the surface, your protagonist might not get *exactly* what she wants in the scene you're about to write. What triggers her misbelief might be the realization that she's getting it for the exact *opposite* reason than she thought she would. This is a very good thing, because it shatters her *internal* expectations, which is what matters most. For instance, imagine nervously going into your boss's office, fully expecting that you're about to get the promotion you've spent years earnestly striving for. And lo and behold, you get it—only to discover that you've gotten it because your boss is positive that you'll fail and then he can give the job to his son without it looking like nepotism. External expectation met, internal expectation crushed.

How do you let the reader in on all this? First, by putting us in your protagonist's head as she approaches the situation. She may be weighing what she hopes will happen against what she expects will happen, and strategizing how to best maximize the odds in her favor. Or, if it's a situation in which she's positive she knows exactly how it will go down (and is, most likely, wrong), she may be thinking about what will happen as if it's already a given. Often, this is what raises readers' empathy, because having been around the block a few times, we're pretty sure that her expectations are going to be dashed.

Ruby's will be shattered. Here's Jennie's answer:

> Ruby expects that the death of Beth's father will not be so bad, because there is so much love in their family, unlike in her own, that it will help them survive the loss and move on.
>
> Uh-oh.

WHAT TO DO

You've been playing along, haven't you? I'm betting that as you read each of Jennie's answers, you were thinking about what your protagonist will go into their origin scene believing. Take a moment now (okay, maybe longer) and jot down your answers to each of the four questions. It's good practice, because this is something you'll want to do before any scene you write.

Writing the Scene Itself

It's now time to write the scene in which your protagonist's expectations will most definitely not be met, and in which his worldview will be skewed. He may emerge emotionally battered and bruised, he may feel triumphant, or he may think that he just dodged a bullet, but as far as he's concerned, life just taught him an important lesson when it comes to navigating the world.

The origin scene will chronicle a single event. It will be specific. You will need to set the place, the time, the context. Don't simply chronicle what happens externally; put us in your story's command center by letting us know what your protagonist is thinking as he reacts, internally, to what happens and to what other people say. Often, what he's thinking and what he's saying out loud will be two very different things. That is the point.

Speaking of what your protagonist is thinking, since origin scenes very often take place when the protagonist is a child, writers sometimes fall into the trap of thinking of "kid logic" as simplistic, surface-level, and on the nose. It's anything but. Kid logic can be more sophisticated than adult logic, because it isn't yet hampered by social custom or euphemisms, nor jaded by world-weary familiarity. Kids see and unabashedly question *everything* precisely because everything is new and unfamiliar. Kid logic is more raw, and so often far more honest than the adult version. It goes just as deep, if not deeper. If you want proof, you have to look no further than the shelves where some of our most beloved literature lives. Scout, the narrator of Harper Lee's *To Kill a Mockingbird*, is six; Jack, the narrator of Emma Donahue's *Room*, is five; and Oskar, the narrator of Jonathan Safran Foer's *Extremely Loud and Incredibly Close*, is the oldest of the lot at nine. All this is to say: don't be afraid of diving deep and giving your protagonist the keen insight of a child making sense of a crisis.

And speaking of diving deep, my advice is to write this scene in the first person, because it's the best way to truly experience the immediacy of what's happening from your protagonist's point of view. In fact, even if you are planning to write your novel in the third person, I would advise you to write *every* backstory scene in the first person—whether it's your protagonist's backstory or that of a secondary character. You can switch back to third person when you start writing the first scene of the novel itself.

Finally, since there will be other people in the scene, take a minute and ask yourself what *everyone's* agenda is. It's never too early to begin asking "why?" of everything anyone does. Just as lawyers never ask a question in court that they don't already know the answer to, don't ask your characters to do anything if you can't answer the question: "Why did they do that? And how does it serve their subjective self-interest?"

You will probably write several drafts before you're satisfied with your scene, because, as we'll explore in later chapters, scenes are envisioned, blueprinted, and written in layers. That's exactly what Jennie did. She wrote this scene about six times before finally arriving at this fully realized version of Ruby's origin scene:

Mr. Anderson was the first person I ever knew who died, and the thing I didn't know about death was how quiet it was going to be. There was this weird hush in the Andersons' house—no one running around, or arguing or throwing balls against the walls. It was as if Beth and all her brothers had been unplugged. Their house now felt like mine did after my sister Nora left for college—a place where it wasn't acceptable to move too fast or speak too loudly or say what you really thought. It was terrifying.

What I couldn't say—what I really thought—was that Mr. Anderson had let me down by dying. We'd made a deal about a month before he died. It happened at their massive farmhouse dinner table on Sunday night, which was spaghetti night. The boys were throwing sourdough rolls at one end of the table, which Mrs. Anderson was trying to referee. At the other, Beth turned to her dad and said, "Ruby's writing a play."

I almost spit out my milk. It had been a secret, for one thing. But for another, at my family's dinner table, anyone who heard that news would just smile and nod, sure I'd never really do it, sure it was just a childish whim, and that would have mortified me.

I curled my toes inside my shoes to brace myself, but Mr. Anderson's eyes lit up. "A play?" he said, and he asked me what it was about, and if he could read it when I finished it. I relaxed, and took a breath. He had no doubt that I would finish it. He had no doubt that it would be worth reading. And for the first time, I felt that confidence, too.

I finished my play a few weeks later. Two days after that, he died. When I found out, I couldn't stop crying—not for Beth and her brothers and her mom, but for me. I was shattered by his death. That's what I couldn't say in their house, because how could I? How could I say how much I was hurt when my friend couldn't even seem to breathe? How could I sit on the couch and cry when Beth and Mrs. Anderson were barely saying a word? I was just the twelve-year-old best friend of the daughter of the guy who had died. My job was just to sit there, eating the cookies the church ladies had put out, pretending that I didn't feel let down.

The church during the funeral was also quiet. I mean, there was organ music and someone sang a sad song about a sparrow, and lots of people spoke, but it still had this strange quiet in the air, as if three hundred people had silently agreed to act as if nothing had changed and the only way we could do that was to make as little noise as possible.

The Andersons all sat together in the front pew, and I kept thinking what it would be like to be Beth and have no father, but all I could think of was that I already kind of felt like I didn't have a father. My dad didn't read to me like Beth's dad had, and he didn't come to watch my soccer games, and even though he was a teacher like Mr. Anderson, he wasn't interested in the stories I wrote or the poems or the plays. Even when he was home he was distant, and I got the feeling he really didn't even like me. I closed my eyes and imagined my father dead, and I didn't feel anything different. A tiny part of me was relieved, because I wouldn't have to pretend to miss him if he died—and that was when I started to cry.

A week after the funeral, our team had a game against the Hornets, and I started begging Beth to play, because I thought that would get her to breathe again and laugh again and be my best friend again. No one loved to kick a soccer ball as much as Beth. "Come play," I said. "It'll feel good."

For the first time since her dad died, she looked me square in the eyes. There was something so dark and sad in her eyes that she looked a little bit dead herself. I swallowed, as if I'd been caught seeing something I wasn't supposed to see. "He came to every game I ever played," she said. "How can you even say that? How can I play without him?"

I understood how much Beth hurt, but she wasn't making any sense. My dad never came to see me play, and I could play just fine. Playing had nothing to do with who was watching. It was something you just did, just you and your teammates. But I wondered if I was missing something. *Was* playing soccer better if your dad came to watch?

"It's worth a try," I said, desperate to save my friend from whatever had taken hold of her.

She shook her head no and looked away again.

Game day arrived, and while we were warming up, I looked across the field and there were Beth and her mom, making their way across the field like they had down the aisle of the church—slow and unsteady, as if each step cost them more than they had to give. I shouted and waved—so excited she was there, she had listened, this would all end—but they didn't look up. Her mom was wearing dark sunglasses and they were all loaded down with stuff—chairs and a cooler. Beth's two younger brothers trailed behind dragging a blanket and a picnic basket. You could picture Mr. Anderson trotting up, swooping in, and helping them all out, but he wasn't there. He was never going to be there again.

Everyone was watching them, but pretending like they weren't watching them, and now that even included me. I hated myself for that.

They came right up to the sidelines where they usually sat. I saw that Beth had on her uniform, but she didn't run out to us. She helped her mom set up one chair and then another right next to it. It was Mr. Anderson's chair, a silent empty memorial.

Her mom quickly sat down without speaking a word to anyone. The brothers sat quietly on a blanket without grabbing for the ball, cracking open the cooler, or wrestling on the grass. Then Beth pulled up her shin guards, turned, and ran out to us, and I thought things would be okay.

The coach called us all into a huddle. We put our arms around each other the way we always did, and suddenly I felt Beth's shoulders shaking. She was crying. Not just quiet tears dripping down her face—she was gasping, howling, and her face looked like it might crack. My heart began to pound because this was my fault, I'd begged her to come, and now everyone was frozen in place watching Beth's guts spill out, and no one else seemed to know what to do to fix it, not even her mom, who still sat, gripping the armrests of her chair as if they might save her from drowning. Beth covered her face with her hands and walked as if in slow motion to the sidelines. She fell on her knees in front of the empty chair, put her forehead on the ground and wailed. "Please, please, please."

She never said please *what,* but she didn't have to. It was please everything.

I had always felt jealous of Beth, but suddenly I felt sorry for her. Suddenly I could see that this was the worst possible thing that could happen to a person, and it had happened to her, and she was only twelve. I could see that she might never recover from this.

Thank goodness, I thought, *I won't ever have to suffer like that.*

Why This Scene Works:

- Did you notice that we instantly start with something that surprises Ruby—how quiet death is? This lets us know immediately that this scene is about expectations that will be broken.

- Did you notice that Ruby looks straight to the past to make sense of the present by accessing a very specific, very revealing memory—the bit about spaghetti night? A memory that also gives us a very clear picture of Beth's home life, not "in general" but in a vivid snapshot that reveals more about their family dynamic than a general explanation would. You can feel it because you are there.

- Did you notice that Ruby compared her own family life to Beth's, providing us with a telling glimpse of the world Ruby lives in, giving us insight into her parents, and letting us know that she has an older sister?

- Did you notice that the memory Ruby has lets us know why she believes what she does about the love she sees in Beth's family?

- Did you notice that two layers of expectations are broken here? We begin with the expectation Ruby had that was broken when Mr. Anderson died—that he would read her play. And, as the scene unfolds, we watch a much more profound inner expectation bite the dust: that the love this family felt would keep them strong, rather than do them in.

- Did you notice that Ruby had a very explicit goal in this scene: to try to put things back the way they were, so that Beth would be her normal self?

- Did you notice that in the end, when Ruby's expectations were not met, her worldview shifted, and she was left feeling glad she would never have what she'd longed for: a closeness like what she'd seen in Beth's family?

WHAT TO DO

Now it's your turn to capture the moment when your protagonist's worldview shifted, and her misbelief took root in her brain, where it's been coloring how she's seen the world from that moment on. Write a full-fledged scene. Don't be worried if it takes several tries to nail it. Feel free to test several scenarios until you hit on the one that feels right. As you saw with Jennie, chances are there will be moments in your own life that will leap to mind, providing evocative material just waiting to be mined. After all, what "write what you know" really means is, write what you know emotionally.

7

WHAT NEXT? THE BEAUTY OF CAUSE AND EFFECT

Cause and effect, means and ends, seeds and fruit cannot be severed; for the effect already blooms in the cause, the end preexists in the means; the fruit in the seed.

—RALPH WALDO EMERSON

When I was six I wanted the same thing we all want: a pony. I wished, hoped, begged, pleaded, and envisioned that pony. He would be black with a cream mane and tail (a combination that doesn't exist in nature, but in my imagination it was lovely). When I close my eyes I can still see him and feel his soft muzzle tickling my palm as he nibbles on the apple I've brought him. But every time I opened the front door, sure he'd be there waiting for me, crickets. Lesson reluctantly learned: You can't simply decide something is true, and bingo, it is.

Having finally accepted that I wasn't getting a pony, I felt giddy, happier, and more carefree than I had in months. Since only a fool would choose to really, really want something they can't have, I immediately forgot all about ponies as if I'd never even seen one. As far as I was concerned, ponies had ceased to be.

Yes, I'm kidding. But unfortunately, there's a pretty widespread belief that once we accept something as true, we will no longer have to struggle with it. It *is* a seductive idea—who wouldn't want to believe that knowledge not only sets us free, but neutralizes the pain and suffering that all that unmet desire would otherwise bring down upon us? However, as most of us have learned through experience, in situations like mine desire doesn't instantly vanish any more than ponies do.

Needless to say, just because I'd realized that my parents weren't going to let me keep a pony in the garage (I mean, there was plenty of space), it didn't stop me from longing for one every minute of every day. Because, not to get too sappy about it, but hey, the heart wants what it wants. And to be very clear, I am *not* talking just about romance here, but about *everything,* whether it's wanting to become a tinker, tailor, soldier, or intergalactic warrior (who is also a spy).

Once we want something, the fact that life says "uh-uh" doesn't snuff out that desire as if it had never been. In fact, it often stokes it. When it comes to your protagonist, over time his desire for the thing he wants might have done a bit of shape shifting; he might have even convinced himself that it's the last thing he'd ever want. Pony? Who needs a stinking pony! He's wrong, of course. Because if that were so, there wouldn't be any internal conflict to stir up, and you wouldn't have a story.

The same is true of his misbelief. You know the fabulous scene you wrote in the last chapter capturing the moment, buried deep in your protagonist's past, when his misbelief kicked in? Well, like his desire, that moment wasn't buried like a rock or an old shoe. It was buried like a seed—one that immediately took root and has been snaking through his life ever since, actively guiding his action, keeping him from getting what he really wants. In the process that misbelief has grown, twisted, and borne fruit.

And yet, it's incredibly tempting to gloss over how your protagonist's desire and his misbelief have affected him since their inception, and instead leap to the start of the story. Which would mean that if, say, at thirteen your protagonist had decided that girls are bad news, then he never once would've been affected by his impetuous resolve to steer clear of the opposite sex from that moment until the novel begins when he's forty- two. When what actually would have happened is that love, pain, and the whole damn thing would have ambushed him at every turn, continually testing, refining, and strengthening his resolve.

So rather than now leapfrogging to the beginning of your novel, in this chapter we're going to spend a little time tracking how your protagonist's misbelief has skewed his life—via three story-specific, conflict-ridden, crossroads moments—keeping him safe (or so he thinks) right up to the second when life (aka your plot) will force him to go after the thing he's always wanted, overcoming his misbelief in the process.

Unlocking Your Plot

Here's the brilliant part about this: writing these scenes will give you potent, specific, and revealing grist for the mill, not only in terms of how your protagonist sees the world, but in the specific memories, ideas, and fantasies that will drive his action, and thus the plot, forward. What's more, these scenes will help establish the cause-and-effect trajectory that will guide your entire novel. And just as important, you'll begin creating the key players—the people in your protagonist's past who, for better or worse, helped facilitate his worldview. Since chances are good these people will play a part in the novel, you'll know when and why they're at cross-purposes with your protagonist, what they're hiding from each other, and when they're woefully misreading each other. In other words, thanks to the specific past you will create for your protagonist, there will already be very specific balls in play when your novel begins, balls that will begin falling to earth, one after the other, creating a hailstorm of unavoidable conflict.

Think of it this way: the plot is karma. Not karma in the metaphysical "what goes around comes around" sense—as in if you're nice to kittens and little children, you will get your dream job—but karma in the very literal cause-and-effect sense, as in if you lie about graduating college, then just as you're about to get that dream job, said lie is bound to surface.

Which means that if you decide that in the past your protagonist lied about graduating college, it's because you *already* suspect that in the future that fact is going to come up, at the worst possible moment, and smack her in the face. By establishing the moments in your protagonist's past that are relevant to the story you're telling, you'll have the material from which to build a solid blueprint. The scenes you'll write will capture moments that have not only actively driven your protagonist's life in the past but are still behind the wheel. In fact, many of these scenes will appear in snippets and as flashbacks in the novel itself. That's why it's crucial to nail them now, and why, even though we haven't actually started developing your plot, this isn't "prewriting," but writing itself.

The other added benefit of this exercise is that your ability to write compelling scenes will also begin to soar. Because by focusing on how what's happening is affecting your protagonist *internally*—what she's thinking, how she makes sense of it, and what then spurs her action— you're mastering one of the most elusive, least-taught facets of storytelling. Ironically, it's also the most important facet—the one that will actually capture your reader's brain. Literally.

What your reader's brain craves is to synchronize with your protagonist's brain as she struggles with a difficult decision, one that will have a clear-cut consequence—that is, a consequence that we can envision and so anticipate. It's virtual reality at its finest, giving us insight into what *we* might do if we were in a similarly difficult situation. And when I say synchronize, I'm talking about a biological fact that was introduced to us by one of the great unsung visionaries of our time.

In Sync: The Truth About Mind Melding

Back in 1966 no one hailed S. Bar-David (aka Shimon Wincelberg) as a prophet, but the *Star Trek* TV writer who came up with the Vulcan mind meld, in which one person's brain syncs with another's, absolutely was. He got everything right. Except, you know, for the Vulcan part. And, okay, the part where you have to put your fingertips on the other guy's face to do it. But hey, the world was pretty much analog back then, so who could blame Wincelberg for seeing life as hands-on, and thus missing the nuances of how information is actually transferred from one brain to another.

To figure that part out, we had to wait for something that even Doc "Bones" McCoy didn't have access to—fMRI technology, which reveals that when we're really engaged in listening to a story, our brain synchronizes with the speaker's brain—literally mirroring it.

In other words, we really *are* on the same wavelength, and their experiences quite literally become ours. The exact same thing is true when we're reading a novel. We *become* the protagonist as our brain waves synchronize with hers, allowing us to viscerally experience what she's going through as she tries to solve the story problem and achieve her driving goal. Spock himself would approve of the logic behind this phenomenon, given that story's evolutionary purpose is to allow us to vicariously navigate unexpected situations from the safety of our own armchair, the better to pick up pointers for surviving them, should they ever befall us on our way to the kitchen for a snack.

Cognitive psychologist and novelist Keith Oatley, who's a tad more familiar with the digital world than Wincelberg was, defines fiction as "a simulation that runs on the software of our minds. And it is a particularly useful simulation because negotiating the social world effectively is extremely tricky, requiring us to weigh up myriad interacting instances of cause and effect."[1]

Remember when we said that a savvy writer relentlessly asks why? Well, the answer to why something happened always takes us back to its underlying cause. In other words, as you are writing specific scenes in your novel, you're continually searching for the real reason your protagonist did what she did, rather than what it looks like on the surface. In real life, all is never as it seems, which is why a story's goal is to uncover what it actually *is*. That is to say, stories reveal the unsuspected cause behind the effect.

And so just as life is ordered by cause and effect, so are your scenes, and so is your entire blueprint. That's why from here on out, cause and effect must underlie everything you envision.

The Beautiful, Irrefutable Logic of Cause and Effect

In our insanely fast-paced, nuanced, multilayered world, where everything is vying for our attention at once, it often seems as if *nothing* makes sense. However, two things remain reassuringly true:

1. There is always a reason for everything—not in the "higher power" preordained sense, but in Newton's humble, simple, hard-and-fast first law of thermodynamics: you can't get something from nothing. In other words, there is *always* a cause. Always an answer to *why?*

2. We turn to stories to find out the *real* reason, the *actual* cause of why something happened, and why someone did what they did. I can't say it too often: Stories are not just entertainment. Stories are the tool we use to navigate life—from your coworker's anecdote about his weird roommate to every article, blog post, and tweet your eyes land on—we plumb stories for information about the people and the world around us. Stories give us inside info into how "if" *causes* "then." As in *if* you're not giving your reader a clear, plausible cause-and-effect progression, illuminated by the why behind it, *then* you're not giving them a story they can follow or care about.

When it comes to blueprinting your novel, the law of cause and effect is one of your most useful tools. Like a very strong flashlight on a very dark night, it reveals the logic behind, well, everything. It's a mathematical proof that you can, with surprising ease, apply to both levels of your novel: the internal story level and the external plot level. On the internal level, the question is, *what would my protagonist's belief/past experience* cause *him to do in this situation?* On the external level, the question is, *how will the other character(s) and the world react to what my protagonist will do?* In other words, cause and effect supplies both the internal and the external logic that underlies and orders your blueprint, ensuring that each event triggers the next.

This is another of those truths that seems so insanely obvious when mentioned that it's hard to believe it could ever be a problem. Ironically, it's precisely because cause and effect is omnipresent that we're often blind to it. It's the reason why, when faced with "what next?" writers often feel like they've parachuted into uncharted territory. They have no idea what lies ahead in their story, or for that matter, the logic behind how they got to where they are now. Even thinking about it can be dizzying, because without a recognizable cause-and-effect progression, nothing makes sense.

By tapping into the supreme (and comforting) logic of cause and effect, that feeling vanishes, giving you wonderfully specific questions to ask, which lead to surprisingly specific answers about what might happen next. This is at the heart of what we will be doing when we start the blueprinting process.

And lest you fear that creating a blueprint based on a clear-cut cause-and-effect trajectory will make your story predictable, worryeth not, because the key word is *might*. A cause-and-effect trajectory doesn't predict what inherently *will* happen; it just lays out the possibilities of what *might* happen. But—and this is the point—it's essential that each one of those possibilities could legitimately be caused by what came before it.

It's this causal "if this, therefore that" link that makes what happens in your novel believable, and that allows your reader to anticipate what might happen next, precisely because they have an idea of what the possibilities actually *are* and what they mean to your protagonist. As no less a literary light than Trey Parker, cocreator of *South Park,* so astutely pointed out when he crashed an astonished and delighted NYU screenwriting class: if your scenes are linked simply by the words "and then," you have something pretty boring; what should follow every beat is either "therefore" or "but." Here he is, riffing to the students: "So you come up with an idea and it's like 'this happens . . . and then, this happens.' No, no, no! It should be 'this happens . . . and *therefore*, this happens.' [or] 'this happens . . . *but* this happens.'"[2] In other words, stories build based on the *causal relationship* between what just happened, and what's about to happen as a result. Makes sense, right?

At this point, it may be tempting to pick up your pen and dash off a summation of the cause-and-effect trajectory leading from the origin of your protagonist's misbelief up to page one. It sounds kind of easy, like something you can do in an hour or so. But you can't.

Why? Because you have nothing *to* sum up. In real life, when you sum up your past, it's natural to take years and years of very specific experiences and summarize them in a sentence or two. After all, who has the time to give a detailed account of each and every experience—and more to the point, who would sit through an endless rendition of it? So we gamely sum it up, and then we draw a strategic conclusion. Like, *Even in the most dire situation, my dad was always aloof, acting as if everything was okay; I thought that was normal, brave even. So I have no idea how to deal with difficult situations other than to pretend they're not there.* But, if pressed, we could then come up with a ton of specific examples of what we *mean* by "dire situations" (the surface event), "acting aloof" (how we read our dad's internal reaction to it), and "I thought that was normal" (the specific, in-the-moment conclusions we drew back in the day, as compared to what we've since realized).

But when it comes to your protagonist, you're trying to summarize, well, nothing at all, because other than the scene you wrote in the last chapter, your protagonist doesn't yet *have* any specific past. Until you've created specific pivotal moments for her, you'll be in the one place we know a writer should never be: the land of the general, with a protagonist whose slate is almost as blank as that of a newborn babe. It's something that in real life would scare the bejesus out of you. Want an example?

Who Am I and How the Hell Did I Get Here?

Imagine waking up in the morning with no idea who you are, what you believe, or what you have to do that day. Okay, okay, maybe that did happen once or twice, in college. But I'm betting that the answers came back to you with unforgiving clarity soon thereafter, along with a splitting headache and a short-lived vow never to imbibe that much again. Now, imagine that happening to you every day—not the imbibing part, but the "who am I and what's my agenda?" part. How frightening it *would* be to start every day with absolutely no idea of, well, anything! It's a feeling S. J. Watson captures brilliantly in his novel *Before I Go to Sleep,* about a woman coping with amnesia:

> I realized I have no ambition. I cannot. All I want is to feel normal. To live like everybody else, with experience building on experience, each day shaping the next. I want to grow, to learn things, and from things . . . I cannot imagine how I will cope when I discover my life is behind me, has already happened, and I have nothing to show for it. No treasure house of recollection, no wealth of experience, no accumulated wisdom to pass on. What are we, if not an accumulation of our memories?[3]

Indeed. And yet writers often begin writing with very little idea of their protagonist's accumulation of specific memories, when in reality, since you can't *really* check your brain at the door, this is the knowledge that their protagonist—just like each of us—has with her at all times. It is the knowledge your protagonist walks onto page one *already in full possession of, and that guides her action from the second she appears.*

The Problem Escalates

So you won't inadvertently give your protagonist amnesia, your goal now is to write three in-depth scenes that helped create, perpetuate, and escalate the problem your protagonist will be forced to deal with when your novel starts. Because the story-specific cause-and-effect trajectory that will propel your novel from the first page to the last doesn't begin on page one; it began with the origin scene you wrote in the last chapter. The first page of your novel, on the other hand, probably takes place somewhere near the middle of said trajectory. We already know how important knowing the protagonist's backstory is. Here's one more thing that writers lose sight of by ignoring it: the protagonist's past is a big part of a novel's force of opposition. Because as we'll see, it tells you what, specifically, your protagonist is up against—both internally and externally.

That's why it's important to trace how the internal battle between your protagonist's desire and her misbelief has affected the story-specific decisions she's made in her life. By the end of this chapter you will have written three turning point scenes in that trajectory (in truth, there very well may be more scenes you'll want to write, or at least identify for later exploration, but let's start with three for now).

These scenes will unfold on a linear timeline, and each one will capture a moment in your protagonist's life when her misbelief was the deciding factor in a major decision she made. Each decision will change the external course of her life, upping the stakes, and be part of the story-specific cause-and-effect trajectory that leads straight to the first page of your novel. None of the decisions she makes will be easy for her. In each scene, she will most likely have a real shot at getting the thing she wants, and *something will happen* that causes her misbelief to rear its head and prevent her from getting it. This in turn will sharpen her desire and strengthen her misbelief, giving it new meaning and expanding its scope and influence on her life.

Step 1: Find the Turning Points

Begin by envisioning your protagonist's life from the origin scene up to the place where—for now—you imagine the novel might start. It's totally okay if you don't know exactly where that is; that's something we'll nail down in the next chapter. But I'm betting that by now you have an inkling of it, and chances are years, probably decades, will have passed between your origin scene and when your novel opens. During that time, your protagonist will have made a lot of turning point decisions that were affected by the battle between her desire and her misbelief. Often simply by focusing on what you already know about your story—and by now you know a lot—you'll find that several possible turning points instantly materialize.

As you sift through the possibilities, keep in mind that you're looking for moments when your protagonist stood at a crossroads in her life and had to make a decision that had escalating ongoing ramifications, rather than random moments that merely exemplify a time when your protagonist acted on her misbelief.

There's no doubt that your protagonist's misbelief has reared its head and made a momentary mess of things more times than she'd like to admit, but most of them are one-offs. For instance, let's say her misbelief is that the nicer someone is to you, the more likely they are to cheat you. That's something that might constantly give her trouble. Like the time the clerk at the dry cleaner noticed that she seemed sad because she'd spilled olive oil on her brand-new jacket, and offered to clean it for free, which made her so suspicious of him that she snatched it out of his hands, stormed out, and ever since has gone to a far less convenient dry cleaner six miles away. That's a drag, for sure, but really, what difference does it make in the greater scheme of her life? None—even though, yes, her misbelief was on full display that day.

Instead, keep an eye out for external turning points that in some fundamental way—via an intense internal conflict—cause the protagonist's misbelief to deepen. Jennie thought about Ruby's backstory and came up with four potential scenes. Here she is evaluating them, beginning with the first idea she had:

> The moment Ruby learns that her sister, Nora, got a puppy. I am thinking that Nora had just moved across the country to go to veterinary school (getting as far away as possible in fact and in spirit as she could from her parents), leaving Ruby alone in an emotionally strained and lonely house.

This moment is big because the girls were never allowed to get a dog and Ruby is totally convinced that Nora will become a fool for this dog— a fool in a way Nora will never again be a fool for Ruby. This leaves Ruby wondering—*if a dog could so easily take my place, did Nora ever really love me at all*?

This idea also made Jennie face head-on something that writers often struggle with: creating a protagonist with a viewpoint that is, as a gazillion posts on Facebook can attest, wildly unpopular.

I am writing about a woman who doesn't like dogs. I know that this is hardly something that seems endearing on the surface, but that's part of the reason I find it interesting. It gives me the chance to explore my own feelings about dogs (and love and loss and grief) by creating a character who grapples with these things in a really dramatic way. Ruby won't— I already know with 100 percent certainty—end up loving the dog. But she will come away with a much better understanding of, and far more empathy for, people who do. I think it's a really fascinating situation to frame a story. And of course it's a major challenge, which is also kind of fun.

Writers often hold themselves back from creating the very protagonist they really want to write about, because they're worried that he or she won't be "likable." So here's something that might come as a big relief: being "likable" is not the point. Instead, you want a protagonist who is relatable, who is flawed, who is vulnerable. Because in a story (and, if we're being brutally honest, in life too) purely likable people are kind of . . . boring. Suspect, even. Someone who's nothing but kind, likable, and perfect tends to leave us wondering *Hmmm, what's he got hidden in the basement, do you think?* And, more to the point, it's not an either/or choice. Neither people nor protagonists are ever just one thing or another. We are all a mash-up of, well, everything.

Okay, back to Jennie's novel. The next three ideas she brainstormed were even more potent than the "Nora gets a dog" scene, and so were the ones she decided to concentrate on:

The moment Ruby first realizes that *not* allowing herself to fall for boys gives her power and keeps her safe. This is a critical scene because I have to show how the Mr. Anderson scene (and her misbelief) impacted her—and right away. So I imagine she will still be young. Say, thirteen. And she will *almost* fall for a boy—someone Beth sets her up with. She'll be following Beth's lead with boys, just to go along and maintain their friendship, and then surprise herself by finding that she actually cares about one of the boys. Her misbelief about love making you weak will be deepened when she holds herself back from falling for him and feels powerful as a result.

This is perfect. It will allow Jennie to probe the first turn of the screw—that is, how Ruby's misbelief began to branch out and direct her action. Then, since the novel will have at its center Ruby's reevaluation of her relationship with Henry, Jennie turned her attention to how Ruby's misbelief affected their relationship from the start. The next scene she brainstormed took place years later:

The moment Ruby meets Henry, she feels for the first time that she's met a kindred spirit. She actually *feels* the thing she has been avoiding all her life—that chemistry, that connection, that pull toward another person—and she lets her guard down, and lets herself fall in love. It's wonderful for a while, but when Ruby glimpses how much she has come to depend on Henry, she has to do something to free herself from the fear of losing him. So, convincing herself that what she felt for Henry was garden-variety lust, and not love at all, she breaks up with him. From now on, Ruby will believe that love and closeness are an illusion—one that she's now too smart to fall for.

This is also good. It will allow Jennie to explore both facets of her novel's third rail: how Ruby's misbelief and her desire interact, each strengthening the other. Remember, like your protagonist, Ruby always has two goals: to find deep human connection (now, specifically, with Henry) and to remain true to her misbelief (that such a connection is too dangerous, and only for fools). Chances are in a scene like this, Ruby's desire would momentarily override her misbelief, causing it to go underground for a bit—gone but not forgotten.

Jennie wanted to keep her focus on Ruby's relationship with Henry, which she knew would continue to have a strong hold on Ruby no matter how much she denied it, so she leapt ahead to what she knew was a major turning point:

> The moment Ruby realizes she can't marry Henry. They've been writing partners for a decade, they love each other, but are both terrified of that reality—Henry because of his parents' bad marriage (I'm winging it here). Their commitmentless relationship has suited both of them for many years, but when Henry finally realizes how different Ruby is from most women (and his mother), he changes his mind and now *wants* to marry her. He asks, she refuses, over and over, until she reluctantly says yes— because he wants it so badly. But as the wedding approaches she begins to feel so vulnerable she can barely breathe. She can't go through with it—she's so stressed she can't even get out of bed—and now Henry feels horrible that he pressured her. So they go back to the way they were— and soon they write this mega hit show. All of which is the ultimate proof to Ruby that her misbelief is, in fact, a solid and correct worldview: you don't need to go ALL IN with love in order to reap its benefits.

Good! A little cliché—but that's fine for now. What works is that all three of these scenes build on each other, as Ruby's misbelief drives her response to her desire, and the action she therefore takes.

WHAT TO DO

Now it's your turn. Your goal is to zero in on three turning point scenes that will yield the most story-specific info, the most potent grist for the mill, so that you can, indeed, begin your novel in medias res. You may come up with many more than three, some of which you'll dismiss out of hand, and others that you may decide to explore in addition to the three you'll pick. That's fine. Just remember that the goal is to have at least three scenes so you can begin to see the escalating arc of your story. Sketch them out the way Jennie did, being as specific as possible. Remember, the more specific you are, the more the scene comes into focus. Specifics play forward; generalities don't.

Step 2: Writing the Scenes

Now you'll transform those three sketches into full-fledged scenes. For an example of exactly how this is done, let's take a look at the first scene in Jennie's three-scene trajectory:

> The thing that made Beth feel better, in the end, was not soccer. It was boys. She figured out that she could use her grief—her tragic aura—to lure them in. The first one was so protective of her, so tender, so sweet, and at first that made her feel worse. She confessed to me that it reminded her of the way her dad treated her mom, and she couldn't stand it, so she dumped the guy—and much to her surprise, she felt a bizarre kind of glee. She was no longer alone in her misery, and as the saying goes, misery loves company. I suddenly understood exactly what that meant.
>
> It was during this time that Beth began to pull away from me. She thought I couldn't possibly understand what she was going through—not with her dad, and not with the boys. I tried going over to her house to do the things we used to do—worm in on her brother's video game tournaments, watch them blow things up in the backyard—but none of the things we used to do were going on anymore. The house was deathly quiet, the boys locked in their separate rooms, her mom asleep on the couch, the refrigerator empty. I was still mad about Mr. Anderson's death, and I still felt sad, but I was still *me*. Grief had completely transformed the Andersons, and I didn't recognize any of them.
>
> It might not have mattered, except that without Beth, I was utterly alone. I was thirteen years old, the only child left in a house with parents who seemed to be paying no attention, a girl who had no idea how to

lure in anyone, male or female, friend or foe. I feared that I lacked some essential quality or skill to connect with people, and I needed to win Beth back to prove to myself that I was going to be okay.

One day Beth stopped to talk to me at lunch. "You should come to the rink tonight," she said.

My heart beat with hope and fear. "Ice skating?" I asked, immediately realizing I sounded like an idiot. Beth didn't go to the ice rink to skate. I'd heard there was a children's party room upstairs that the kids snuck into. People whispered about what went on in there, but it was all vague and shadowy to me. I had no real idea, which made me feel foolish on top of everything else.

She laughed. "No," she said. "Brian has a friend who thinks you're cute."

"Okay," I said—not thinking about what I would have to do with the boy, or how I would have to lie to my parents about where I was going, but thinking only of getting Beth back.

When I stepped into the dark rink, the entire building seemed to thrum with energy and the beat of Michael Jackson. Even though it was cold, I was sweating. Beth had told me to sneak through the door leading upstairs when the greasy guy at the snack bar was scooping popcorn. I waited for my moment, then slipped in.

There were couples lying together on the couches and against the walls in the dark. I stood there, frozen with fear—was Beth having sex? Did she expect me to have sex?—when suddenly someone caught my arm.

"Ruby?" a boy asked. "Beth's over there." This must have been the friend who had singled me out.

I peered into the corner and saw Beth sitting on the lap of a boy who had to be sixteen or seventeen. Her skirt was hiked up to her hips, but she was still dressed. Everyone was dressed, but instead of feeling relieved, I panicked, turned, and bolted back into the hallway.

The boy followed me, fast on my heels. "Leave me alone," I said.

He grabbed my shoulder from behind, spun me around, and pinned me against the wall. I squeezed my eyes shut, bracing for some kind of blow. "Hey," he said, "what's the matter?"

I pressed my lips together but couldn't stop the tears from filling my eyes and spilling down my face. I expected him to leave then, but instead he said, "Come with me," and he took my hand and pulled me through the crowd. He shepherded me downstairs and out the doors into the dark of the night. He pointed to a bench just out of the pool of light, and we sat—his hand still wrapped around mine. He didn't move. He didn't speak. And gradually, my heart stopped slamming against my chest.

We sat like that for a while, in silence, but it was a different kind of silence than I'd ever known—not the hollow silence of my house or the guarded silence of Beth's. It was a comfortable silence, a safe silence. That's when he leaned over and gently kissed me.

In an instant, I felt a crushing wave of dark fear—because out of the blue, out of nowhere, without any warning, I had something to lose. This boy, whose name I didn't even know. All I had to do is squeeze his hand, smile at him—and that's when I knew exactly what I had to do to get Beth back.

I forced myself to let go of his hand. I made myself stand up. "I have to go," I said.

His face was twisted in confusion. "No, don't," he said.

I stepped back toward the door. Just as I was reaching to open it, Beth pushed through—as if on cue. She looked at the boy, who sat on the bench still wearing a look of hopeless longing, and she looked at me. I grinned—the sly grin of the victor—and she grinned back.

I knew then we'd be back at the rink tomorrow night, and that if this boy were there, I'd walk right past him as if we'd never met.

Why This Scene Works:

- Did you notice how Ruby used the memory of Mr. Anderson's death to make sense of what's going on here?
- Did you notice that Ruby was constantly conflicted? She wanted to feel connection with someone, and the only person she had left was Beth. So she decided to do whatever it took to reconnect with her, even if it was something she wouldn't normally want to do.

- Did you notice that her decision scared her? Fear sits right there next to longing.
- Did you notice that when she suddenly achieved the connection she was looking for—not with Beth, but with the boy—*that* frightened her more than anything? And again, in this she referenced the past and what it taught her.
- Did you notice that in the end her misbelief was strengthened *and* she got what she thought she wanted: connection with Beth, which of course, was no real connection at all?
- Did you notice what else she got in that moment? She got a taste of what genuine connection actually feels like—which will surely help fuel her desire for connection.
- And finally, can you see how this new experience might drive her behavior for a long time to come?

WHAT TO DO

Now it's your turn—and again, don't worry about getting it right. Start with the first of your three turning points, and write each scene in chronological order. You've already been to this rodeo; you earned your spurs when you wrote your origin scene. That scene is going to come in handy now. Take a minute and reread it. Put yourself in your protagonist's brain, see the world through her eyes. Now, turn those eyes toward the three scenes you're about to write. Aim for scenes as fully fleshed out as the one you just saw Jennie write. This might take awhile, so have patience with yourself, and give yourself permission to take as long as you need to bring each scene to life. The good news is that the scenes do not have to be polished, or "beautifully" written. They just have to capture these escalating turning points in your protagonist's life. Nailing them down is well worth it, because when you're finished, you'll be ready to ask the question that uninformed novelists often start with, much to their peril: Where does my novel begin?

8

THE WHEN: AN OFFER YOUR PROTAGONIST CAN'T REFUSE (BUT PROBABLY WANTS TO)

Everything must have a beginning . . . and that beginning must be linked to something that went before.

—MARY SHELLEY

We've spent the last five chapters digging, working on the first half of your story, so that your novel can, indeed, start in medias res. You've created the bones of the first half of the thing, and it's almost time to think about the thing itself. We have one last question to answer before we arrive at the first page. Because now, in a remarkably short time, we've arrived at what's often one of the most elusive things to pin down: the "when"—as in when *does* your novel start?

The simple answer is that it starts when life will no longer allow your protagonist to put off going after that thing he's long wanted, regardless of how much his misbelief—and, as we'll see, his biology—suggest he sit this one out. Because no matter how dearly we want something, avoiding change is our middle name. That's probably why the only thing that causes us to change, internally or otherwise, is an unavoidable *external* force. In other words, a problem that we can't dodge, duck, or deflect is barreling straight toward us, giving us no choice but to take action.

Welcome to the underlying conflict that fuels all effective narratives: story is about change, and we're wired to avoid change. Ask us to change, and we reach for the "opt out" button. Push us a little harder, and we instinctively dig our heels in. Think of our response as the evolutionary equivalent of that schoolyard retort: *Oh yeah? Make me!* Let's face it, if we have a choice, when the going gets tough, we'd probably get a snack, turn on a movie, and hope it all blows over. Who wouldn't? But here's the interesting thing: that movie we then turn on? It's going to be full of exactly what we're watching it to avoid: risk, danger, internal conflict, change, and, we hope, a hard-won reward at the end.

It seems ironic that the very things we give a wide berth to in real life are exactly what we crave in stories. But it's not ironic at all. It's a big part of why story evolved in the first place: so we can vicariously experience the cost, and reap the benefit, of the changes we so studiously avoid in real life. Think of it as a dress rehearsal, so when life forces us out of our well-established comfort zone, heart pounding, palms sweating, we can bravely venture forth into unfamiliar territory. Stories help create the internal compass we use to navigate the unexpected changes that life is so fond of tossing in our path, often at the worst possible moment. And when it comes to your protagonist, the question, of course, is what *is* the worst possible moment?

To that end, in this chapter we'll explore why change, even good change, is so damn hard. We'll make sure that your novel's overarching plot problem has a clear consequence that the reader can begin to anticipate from page one, with a ticking clock that's capable of sustaining an entire novel. And finally, we'll be sure that said plot problem is firmly harnessed to your novel's third rail, so every twist and turn will force your protagonist to struggle mightily. For his own good, of course.

The Surprisingly Reassuring Beauty of the Unavoidable

Up to now, while life may have slammed up against your protagonist's misbelief, there was always an out, a way for her to resist, and so to stay true to it. But as your novel opens, there's one thing your protagonist's misbelief hasn't completely crushed: her desire for that thing she wants. She's probably done a pretty good job of sublimating it, allowing her fear and her desire to live side by side in an uneasy truce of sorts. She's an ace at avoiding the conflict she'd face if she ever really stood up to her misbelief. And so just before your novel begins, chances are she's either congratulated herself because her life is exactly the way she wants it—or, a little more realistically, she's acknowledged that life sucks, but since there's absolutely nothing she can do about it, she's made peace with it. Thus she's floating in the middle of the ocean, settled down in her customary boat, and the water, though perhaps not calm, is predictable, familiar. She doesn't have to worry about sinking or swimming. As she bobs along, unaware of how incredibly vulnerable she is, you almost feel bad for her. Because you know that a torpedo is being primed to ram into her unprotected little vessel, upending her routine and tossing her into water teeming with sharks.

That's how stories begin. The protagonist thinks everything is on course—and then, *bam!* Life says, "Think again." That is the function of plot. Something happens that forces your protagonist's hand, leaving her no choice but to take action. And that, as it turns out, is a very good thing. Which is precisely the beauty of problems we can't avoid: they force us to make the changes we've always wanted to but—okay, let's be totally honest here—were too chicken to try.

JFK said it best. When asked how he became a war hero, he grinned and replied, "It was involuntary. They sank my boat."[1]

No pain, no gain; it really is as simple as that.

And speaking of pain—and keeping in mind that all change is hard, even good change—it's time to switch gears in terms of how you treat your protagonist. Thus far you've lovingly created your protagonist, empathizing with him as each twist of fate further cemented his misbelief. When that misbelief then caused him to make questionable decisions, chances are your heart went out to him. Because you understood his true motivation, which was probably very different from what it looked like on the surface. Now, you have to steel yourself, because you're about to start crafting a plot that's going to pull the rug out from under him when he's at his most vulnerable. It's time for life to hit him with its best shot, or, to paraphrase the no-nonsense wisdom of Don Corleone, to make him an offer he can't refuse (hopefully sans the horse head).

This is where many writers stumble. You're a nice person, plus you've become so fond of your protagonist that you don't want to cause him pain, especially not on purpose. So instead of pulling off the gloves, you're sorely tempted to begin pulling punches. You want to be *fair* to him. You want to give him the benefit of the doubt. You don't want to judge him. But here's the thing: we're not talking about you. We're talking about life, aka the plot. And life isn't fair. That's why we need stories—to figure out how to deal with all those unfair things that happen, so *we* can have the strength and the wisdom to be fair ourselves. If life doesn't pummel your protagonist hard, he can't figure out what's fair and what isn't, let alone muster the courage, moxie, heart, and smarts to survive in a world that can be so darn unfair. In other words, not only won't he have anything to teach us, but he won't be a worthy teacher. So fasten your seatbelt, as the fabulous Margo Channing said in *All About Eve*; you're in for a bumpy night. And so, I hope, is your protagonist.

Why *Is* Change So Damn Hard?

Why *are* we such masters at forever putting off 'til tomorrow what, apparently, we could have done today? Especially since in hindsight, we almost always wish we *had* done it yesterday. Are we simply stubborn? Lazy, maybe? Or worse, cowards? While each of those things might be true in certain specific instances (not that we'd admit it out loud, mind you), they're not the reason. It turns out our resistance is not a personal failing at all, which is probably something of a relief (it sure was for me). Instead, we're hardwired to fight change, often at all costs. Our basic resistance is, in fact, a long-standing survival mechanism. Evolutionarily speaking, in almost every instance, resisting change didn't make us stubborn, it made us smart. Note the *almost*.

Here's the skinny. Voluntary change—that is, change not caused by a compelling outside force—runs contrary to our most basic biological directive: to survive. Not survive in luxury, let alone comfort, but survive as in live to see the dawn. As long as you're surviving, even if you're miserable, you're safe. So why rock the boat?

Safety first, change second. And that *almost*? Since there are times when change is not only a good idea, but essential, there is a scenario that overrides our resistance by rendering it moot. That is when the boat seems to start rocking all by itself, giving us no choice but to leap into the water. In other words, change is spurred by outside forces that can't be avoided. Otherwise, we're more than happy to put it off for one more day. This is most likely what accounts for our sometimes baffling propensity to hold onto bad habits even when we know they're not serving us well. So you can stop blaming yourself for the fact that you're still driving the clunker you swore you'd sell last winter, still eating meatloaf on Thursdays even though you don't really like it, and still haven't gone to the gym once this month (okay, year). It's not you. It's your biology.

The technical term for the reason we're so averse to change is homeostasis (a nifty word that might come in handy when doing the Sunday crossword puzzle). Basically homeostasis means that once a system's in balance it tends to stay in *that* balance, because experience has proven that it's safe. So when a biological organism, be it an amoeba, a yellow-bellied sapsucker, or a human being, finds an ecosystem that ensures its physical survival—think food, water, temperature, shelter—its natural reaction is to hunker down and stand as pat as a blackjack player who's been dealt two kings. In the survival game, if what you've done so far has kept you breathing, then you're holding a winning hand. Provided, that is, that you're living in the rough-and-tumble world in which we evolved—and therein lies the rub.

You're not.

Until very recently humans were as much a part of the food chain as free-range chickens and quinoa are today, meaning we had to watch our backs, pretty much 24/7. It was the trials and tribulations of the early Stone Age that our brain evolved to deal with, and that our wiring still works hard to keep us safe from. Evolutionarily speaking, the word "safe" merely means safe from imminent physical danger. And back then, to *be* safe from imminent danger was a major accomplishment. Yabba dabba do!

Since then, we've created a world that's quite different from Fred Flintstone's, but our basic biology didn't get the memo. It remains attuned to the world we evolved to survive in—one in which from dawn to dusk we foraged for food, water, and shelter, all the while keeping a sharp lookout for famished dire wolves. The fact that our original wiring is still in operation makes sense, given that the Stone Age world stayed pretty constant for eons, while evolutionarily speaking our modern world is about five seconds old.

But here's the catch: along with our human-made world (water from a tap, food from the market, shelter based on your credit score), our concept of safety has evolved, too. Just breathing no longer cuts it. We don't just want to survive physically, we also want to be happy doing it. We want to feel fulfilled, purposeful, and, let's face it, comfortable, as in "I have a nice house, a happy family, and a bit of money in the bank." That's the modern definition of safe. But doing what it takes to achieve that kind of safety doesn't have the same inevitable sense of urgency as does hearing someone yell, *"Lion! Run!"* I mean, you can always quit that dead-end job tomorrow, or the next day, or, let's be honest, never, and still survive to see any number of dawns. The point is, we're still maddeningly hardwired to resist voluntarily making the very changes it often takes to get what we want. Enter story: our go-to method for learning to spot, and navigate, exceptions to the "stand pat" rule, so we can happily ride off into the sunset in the manner to which we've become accustomed.

That is why, as your novel begins, your protagonist has most likely spent a good bit of time downplaying, postponing, and often willfully ignoring the urge to change. In other words, he's rationalizing—sometimes consciously, but more often than not, as far as he's concerned, he's simply making strategic sense of the world, and acting accordingly.

Which brings us to one last stumbling block when it comes to making voluntary changes: our ability to look on the bright side, even when it's pretty darn dark. While that tendency can get us through some very stormy nights, and we often make thirst-quenching lemonade from all those lemons, there are times when pretending things are fine when they're not is actually the very thing that does us in. Why? Because we get so used to things not being fine, that we forget what fine is, and so what we once recognized as a problem no longer seems like a problem at all. That's not good, because not only does it remain a problem, but it's now free to poke around, unobserved, and uninhibited. Thus it gathers force just beneath the surface, until at long last it erupts, blowing all our rationalizations to kingdom come.

It's the same with your protagonist. That moment—the one when the problem finally has the firepower to override his ability to ignore it—tends to be when your novel begins. That moment is exactly what we will pin down by the end of this chapter. But before we can do that, we first need to clearly identify the overarching external problem that your protagonist will battle from the first page to the last. The first step is to examine what you know so far about your protagonist's predicament.

Brainstorming Where to Start: What's the Problem?

The first question to ask about your protagonist's predicament is, what unavoidable external change will catapult my protagonist into the fray, triggering her internal battle? In other words, what threshold is my protagonist standing on the brink of, whether she knows it or not? Who (or what) is taking aim at her little boat? Keep in mind she might be the culprit herself—in fact, self-sabotage can be far more ruthless, not to mention infinitely more effective, than the third-party kind that conspiracy theorists have always been so fond of yammering about on short-wave radios from bunkers in the badlands.

Given the work you've done thus far, you probably have one or two opening scenarios in mind. Sometimes the overarching plot problem is blazingly obvious: in *Gone Girl* it's that Nick must deal with Amy's unexpected disappearance; in *Lord of the Rings* it's that the one ring to bind them all is getting antsy and about to do just that; in *Jaws* and *Moby Dick* it's a fish (okay, a fish and a mammal that's very fishlike).

But what do you do if there isn't an obvious external problem on the horizon? Or what if it seems like there are a lot of problems, but none stand out as *the* problem? For instance, Jennie's story is about how Ruby realizes that you can't hide from love, nor can you hedge against it, and so despite all her efforts to pretend otherwise, she actually *has* it with Henry, but hasn't let herself (or him) really sink into it and savor it. Nor has she formed a tight bond with anyone else: in fact, an entire layer of her life has been stunted. Ruby will have to find a way to make peace with that, hopefully growing in the process, and finding the strength to go forward.

Okay, so what's the plot then? What will happen externally that will force Ruby to make that internal change? What *actual* external problem is she facing?

There's nothing in Jennie's description so far that gives us much of a clue, because right now it still lives in the land of the general. Jennie knows that grief is going to play a big part, but grief over what, exactly? Going all the way back to her *What If,* there was something about Ruby's career being threatened, and that she was going to have one chance to make things right. But as we wondered back then, make *what* right? Right now when it comes to specifics, all Jennie knows is that Henry is going to play a big part in it. You may be facing a similar quandary. If so, don't worry; given all the digging you've done, chances are several potential problems are hiding in plain sight.

Here's Jennie working it out:

> Okay, so Ruby's going to hurt—badly—as a result of the love she thought she'd protected herself from. She'll be so blindsided that for a while even breathing will be a challenge. So something's got to happen with Henry. He could fall in love with someone else, sleep with someone else, write with someone else, maybe die. I am completely not interested in writing a story with sexual betrayal at the center—just zero heat for me in that

idea. So either Henry decides to end things with Ruby, or something pretty bad has to befall him. And, because Ruby's career is in question too, maybe this occurs at the moment when something important is happening with their show, something that puts Ruby's credibility into question. That would put her in a dire situation and also force her to prove—um, ha! I don't know. To prove she really was an integral part of the writing team, to prove she really did love Henry, to prove she wasn't a total heartless jerk?

Notice how Jennie has focused in on potential external events geared to stoke the inner struggle she wants Ruby to go through? Her goal now is to find the main problem. The one that will not only escalate itself, but cause the other problems to build in unison. To do that, she made a list of all the problems that her general scenario suggested:

- Henry decides to write with someone else.
- Henry gets too sick to write.
- Henry actually dies.
- Something to do with the show they're writing, a deadline maybe, that Ruby must meet alone.
- Hey, it *is* L.A., so maybe there's an earthquake, the whole city comes to a grinding halt, and the show get preempted.

When Jennie reread her list, she pulled up short.

Wait—it's odd there's not one thing about a dog on this list. My whole plot is going to revolve around Ruby stealing this dog, and the dog turning out to be not quite what she thought. Since she's in TV, and as I just pointed out this is L.A., I keep thinking about a famous person's dog, a dog the whole world quickly mobilizes to search for. Ruby steals Johnny Depp's dog—that kind of thing. The way I've been picturing it, someone will recognize the dog almost immediately, and so Ruby will be on the run with the dog for most of the novel. Shouldn't that be on this list?

The answer is no—but in asking the question, Jennie has just pinpointed a major inadvertent error writers often make: mistaking the events in the plot for what the story is about. What we are doing here is trying to nail down the story, which, as we've seen, is very different than the plot. The dog stuff belongs in the realm of the plot.

WHAT TO DO

Do a little free writing about your intended plot the way Jennie did. Then extract from that a list of as many ideas for your novel's main problem as possible: the ones you already had a notion about, perhaps new ones that are just occurring to you now, even ones that seem far-fetched. Don't worry about sorting them out; the goal is simply to identify as many as you can. Sure, some of them will ultimately be discarded, but it's not just the main story problem you're identifying here; it's also secondary problems that it will bring to the surface and drive forward. So none of this effort will be wasted.

Putting Potential Plot Problems to the Test

Now that you have your list, how *do* you choose the main story problem? It's of the utmost importance, because as we know, a novel is about *one* problem that complicates everything else. You already know what your novel's internal problem is—that is, the third rail. Now you're homing in on the one *external* problem, which will get its ultimate power by coming into contact with said third rail.

Luckily, there are two tests that will enable you to gauge each problem's potential. The first test is external; the second, internal.

Plot Problem Test 1: Can the Problem Sustain the Entire Novel from the First Page to the Last?

This test has three hurdles that each problem must clear in order to stay in the running.

Can the problem build?

The problem that kicks into gear on page one must have the stamina to play through your entire novel, sparking the third rail and picking up speed as it thunders forward. That's a tall order, and why you can't pick just any old external problem.

To be very clear, this does not mean that something massive has to be happening on page one. The question isn't whether your problem is big enough—a tidal wave, an earthquake, a hurtling meteor—but whether your problem has the power to grow, intensify, and complicate. Legendary movie mogul Samuel Goldwyn is said to have advised: "What we want is a story that starts with an earthquake and builds to a climax." In other words, even if it starts big, the real question is, can it sustain *growing* momentum? That's why what's happening in your opening scene can't be something that merely dashes your protagonist's expectations *today,* and why we are working on the overarching problem first.

Because just like your dog, your beloved, and—it's so hard to admit this—your love handles, the problem that kicks off your novel must be in it for the long haul. Not any old problem will do. And certainly not one that's easily solvable—although on page one it might *seem* like a cake-walk to your protagonist. In fact, it probably will. In reality, however, that problem must be the tip of an iceberg your protagonist has mistaken for a couple of ice cubes. Or, with apologies to Robert Frost, for those who favor fire, it's not about being able to put out a slew of small fires, one after the other—rather, it's about trying to put out one small fire that turns out to be far more potent than it first appeared. So sure, a novel can start with a small fire. Novels often do. But here's the thing: a story is about how, in trying to put out a seemingly minor blaze, the protagonist inadvertently fans the flames, until by the end, it's a raging inferno.

It sounds so reassuringly logical that it's a little surprising how often writers instead create a plot made up of a series of separate obstacles, each aimed at tackling an individual facet of the protagonist's struggle. This approach causes readers to bail in droves, or it would, if there were droves to begin with. Here's why:

- Every time the protagonist solves a problem it feels like the novel has ended. The momentum stops because the reader now has nothing to anticipate, and so no reason to read forward.
- Each obstacle tends to carry the same approximate weight and meaning, so it begins to feel like, second verse, same as the first. And the third, and the fourth.
- Although the writer has painstakingly designed each obstacle to test a certain part of the protagonist's mettle, the reader never makes that connection, because it's conceptual. Plus, what the protagonist learns from having solved the first problem has no effect on what she must do to solve the second, so the reader starts to wonder what the point is.

And when writing such a novel, it's not long before the *writer* begins wondering the same thing. There's no more disheartening feeling than having no idea where to go next. That's a trap you will deftly avoid by harnessing your novel to one escalating problem. It will provide the foolproof yardstick we've been talking about, allowing you to gauge what's relevant, and just as important, what isn't. It's what will stop you from arbitrarily throwing in random quirks, dramatic events, and irrelevant bits of backstory to spice things up, including adding a character because Aunt Margaret thinks it would be hilarious if the next-door neighbor always told dumb jokes just like Uncle Bill does (it would serve him right to see himself in print, too).

So look over your list and cross out any problem that falls short of the mark because it's too easily resolved, or not specific enough to really challenge your protagonist in a meaningful (and hopefully painful) way.

When Jennie examined her potential problems, she realized that all of them were still in the running except, taking Goldwyn's point, the earthquake idea. It's a temporary problem, and one that doesn't lead anywhere. But there's more to an evolving plot problem than staying power. It also has to *be* a problem, and problems have consequences. So for all those dilemmas that you've determined can play out over the course of your novel, the next hurdle to clear is the following.

Is there a real-world, specific, *impending* consequence that this escalating problem will give my protagonist no choice but to face?

There must be something clear and definite that will occur if the protagonist fails or, worse, doesn't take action. It can't be vague, conceptual, or iffy.

When Jennie saw this question, her first thought was, "Are you sure I can't add the dognapping scenario to the list? It would pass this test!" She has a point. After all, Ruby's going to snatch a pooch the whole world will soon be looking for, and the impending consequence is crystal clear: at some point in the future either they'll find the dog or Ruby will give it up, and she'll have to face the music.

True. But that's not the kind of impending consequence we're talking about. Why? Because although it is a real-world ramification, it centers on a random plot point, instead of the story itself. What does it have to do with Ruby's grief, and her struggle to overcome her misbelief about love and human connection? What does it have to do with the third rail? Not much.

So Jennie thought about it again, and this time she focused on the most dramatic event on her list: Henry's sudden death. That would last forever (talk about staying power!) and catapult Ruby into an unavoidable ongoing problem: she'd be forced to live without the one person she's closest to, changing her life forever. There would no doubt be lots of impending consequences that Ruby would have no choice but to face. So for a moment it seemed perfect. Until Jennie crashed into the third and last hurdle:

Is there a clear-cut deadline, a ticking clock counting down to that consequence?

Once again, Jennie realized that what looked like the perfect answer, wasn't. Because if Henry is already dead when the novel begins his, um, deadline has already passed. And so while there are a slew of potential consequences, there is no one specific thing that Ruby has to do, and no time frame to do it in. "Okay," Jennie said, "so what if the novel opens with Henry on his deathbed in the hospital, his time short—say, a week at most?"

Nope, not there yet. Because as Jennie quickly realized, regardless how painful the consequences when Henry dies, there's nothing specific hanging in the balance, no quest, no problem that his impending death forces Ruby to solve. The only clear consequence from beginning to end would be that Ruby will be very, very sad. She doesn't have to do anything at all. She doesn't have to *act*. A key rule of thumb is this: if at any point your protagonist can simply decide to give up without suffering great personal cost *due to her inaction,* you do not have a story. In this scenario, Ruby has that choice, so it's a no-go.

With that in mind, Jennie went back to her list in search of a plot problem that would clear all three hurdles. She jettisoned the thought of Henry deciding to write with someone else, knowing that his death would be the only thing capable of causing the grief she's always known Ruby would feel. That led Jennie to the one problem on her list that might force Ruby to act even though she's overwhelmed with grief: a writing deadline that she must meet. Here she is working it out:

> So going back to my initial sketch of Ruby, what if a few months before the novel begins, Henry decided that their hit show must come to a close because he has taken a big screenwriting job—without Ruby. She's reeling from the blow, and from the fact that she'll lose their daily intimate writing life together. As a result of the tension between them, the writing of the final season is rough.
>
> A week before the action starts, the cast and crew are right in the middle of filming the final episode, when the studio head, the producer of the show's boss, insists they rewrite the script. They have a drop-dead deadline of ten days. Henry and Ruby argue about how to rewrite the ending—and because of their argument, something happens to Henry; an accident, let's say. He's unable to write. Maybe he's in a coma, about to die.
>
> Now Ruby has to deal with Henry's impending death and the fact that she has a very short time to rewrite the finale—alone. And maybe writing it isn't just about finishing the series, but about *what* she writes, how she ends it. This could be where Ruby reveals things to herself

and to Henry, and maybe even the whole world, about who she really is. Maybe there are external threats, as well; maybe there are die-hard fans waiting in the wings eager to revise the finale should Ruby go to pieces. So even though she's gobsmacked with grief, this is her only chance to set things right. The problem is, she's so undone she can't write a word.

Yes! This is an external quest that Ruby cannot avoid without great personal cost. You can just *see* all the richness of that coming into play here. Sure, Ruby could still choose to sit there and do nothing, but now her inaction would have major consequences that would impact her big-time.

Like what?

It's a great question, and one that Jennie can begin to answer because it's specific. Just thinking about it begins to yield juicy possibilities: Ruby would have to deal with the fact that she's letting down her fans and the people who work on the show. Plus, maybe an irritating fangirl will offer to write the ending, which would drive Ruby crazy, especially since her inability to write might lead fans *and* colleagues to amp up their speculation that Henry was the show's creative force, and Ruby just a tagalong. So Ruby could lose her reputation and her claim to her greatest accomplishment, and never be hired to work again. In other words, there are lots of great, story-worthy specifics ripe for development.

So in the end, although most of the problems Jennie considered had a deadline, only one had the power to force Ruby to take immediate action that's directly related to the story she's telling. And here's something interesting—this effective overarching problem actually incorporated a great many of the original ideas Jennie had for her story. Now, before the action begins, although Henry doesn't find another writing partner, he decides to write solo. And then something happens that will, in fact, cause his impending death. Thus none of what Jennie started with has yet been totally discarded. Except for the earthquake.

WHAT TO DO

Run your list of potential plot problems through Test 1. Be ruthless; don't let any problem through unless it clears all three hurdles. Chances are, like Jennie, your problems have begun to evolve and merge. Even so, you may have several still in the running. If so, get ready, because the dilemmas that have passed the first test—the external test—now have to face the internal test.

Plot Problem Test 2: Is the Problem Capable of Forcing Your Protagonist to Make the Inner Change That Your Novel Is Actually About?

While each of your potential plot problems might have a ticking clock, is the clock you're considering the *right* one? Meaning, does the ticking clock represent a plot problem that will continually touch your novel's third rail? This is no small consideration, because that's what gives your external problem the strength and power to drive your novel home. To determine which plot problem is the most connected to your third rail, you'll need to answer these two questions.

Will the problem's impending consequence force my protagonist to struggle with her misbelief?

"Arggh!" (That's Jennie.)

The answer at this point is no. At least, not yet. While there's clearly the potential for the script scenario to touch the third rail here and there, there's not a direct, continuous connection strong enough to power the novel. It's just something that Ruby has to do because it's happening in the here and now, and because her writing partner was in an accident. So it's largely external.

Here's Jennie working out how she can harness that scenario more forcefully to the third rail:

What if the series, along with everything Ruby and Henry have written together, is where Ruby plays out the life she mocks in real life—a life built around genuine connection. It's actually her fantasy writ large. And, because she writes these shows with Henry, it's also where their closeness is in full force. So, unbeknownst to her, she is actually experiencing in her work life the very thing she works so hard to avoid in her personal life. In the novel, she will have to face all this because she can't simply walk away from the show. The ending has to be rewritten—and she realizes that the only way it will work is if she has the courage to write about the kind of love she has been dismissing and avoiding all along. The show must go on—and she must face her misbelief.

Bingo! Jennie just found the deep connection between her overarching plot problem and her third rail. Rewriting the show's finale will force Ruby to face her misbelief head-on, which terrifies her.

There's just one more question to answer.

Regardless of whether or not my protagonist achieves his goal, will the approaching consequence cost him something big— emotionally speaking, that is?

At long last, the answer for Ruby is *absolutely!* And the emotional cost will then do exactly what it's meant to do in every novel: shatter the protagonist's worldview once and for all. Remember, your plot problem must have the power to ruthlessly spur your protagonist to change, or (figuratively at the very least) die trying.

Done! Jennie now knows what her novel's overarching plot problem is: Will Ruby revise the finale on time? That is the main ticking clock, and that is what we can use to identify where her story should start.

Here is the secret: Your novel's main ticking clock is what all the other clocks in your story will be set to; it's what will give them their meaning. Every other problem you identified must be able to spin off of, inform, and complicate the overarching plot problem your novel revolves around, affecting the resolution it's barreling toward. In Jennie's novel, Henry's

looming demise, the annoying fangirl who wants to write the finale herself, and the hunt for the dog Ruby only intended to borrow for an hour
or two, will do just that.

WHAT TO DO

Run your remaining plot problems through Test 2, until you've winnowed
them down to a single, overarching dilemma that touches your story's
third rail. It often takes a bit more digging, refining, and tweaking, as
Jennie's story did, to get there. They may seem like minor nitpicky tweaks,
but they're critical tweaks. Don't skip over them!

Keep at it until you've developed one overarching plot problem that
leaps over every hurdle and meets both tests.

Look, Up Ahead; I Think I See an Opening!

We've finally arrived at the seminal question: When does your novel
begin? What, specifically, will happen to start the chain reaction that will
cause *everything* to happen? What will trigger your protagonist's decision
to take that first step out of her comfort zone? The good news is, your novel's ticking clock will lead you directly to the answer. It's simply a matter of
zeroing in on that seminal tick. And just like in real life, it's never the first
tick. You rarely even notice the first tick as anything out of the ordinary.
It's not until the fourth or fifth tick that the sound breaks into your con-
sciousness and it dawns on you that something just might be wrong—and
by then, the problem has usually grown considerably. Let's watch as Jennie
tracks the first tick of her overarching story problem up to the telltale tick
that will force Ruby to do more than sit in her house and ruminate.

First Tick: Henry and Ruby write their final script. Normally, they write
very well together—they had developed a great system, where Ruby
throws out ideas, Henry questions and pushes and prods to add meat

to the bone, and then Ruby sits down to actually put the words on the page. They had this *down*. This time, however, it is a rocky process. They know it's their last script. Henry feels guilty for leaving Ruby to write the movie (and is probably nervous about whether he can do it without her). Ruby feels abandoned and angry and isn't shy about making her feelings known. They don't agree on how the story should end, and Ruby ultimately agrees to Henry's ending, but phones it in. She's amazed when their producer—let's call her Sharon—puts the script into production.

Second Tick: In the middle of filming, someone at the studio asks for a last-minute rewrite of the ending—making Sharon look bad, putting huge pressure on Ruby and Henry, and reopening the fresh wound. Ruby and Henry argue about how to proceed.

Third Tick: Because of the argument, Henry has a tragic accident— I'm thinking he gets hit by a car, maybe while riding a bike (my husband is a cyclist and that terror is ever present). Everything is a blur as the doctors determine his prognosis and Ruby tries to absorb the news.

Fourth Tick: By the time it's clear that Henry isn't long for this world, Ruby has less than a week to finish the script. She can't breathe, let alone write. She's deep in grief and denial.

Fifth Tick: But then something with Sharon and the fans forces her to take action—I am thinking that a fangirl, let's call her Clementine, has written a script that the producer thinks is actually a viable replacement— and there is no way Ruby is going to allow that to happen, so she resolves to pound out the finale. But she's never written a script on her own, let alone that quickly, and she has the whole problem of *how* to end it—the way Henry wanted, or the way she thought would be best. She's so lost that she can't do much of anything; she's not eating, not sleeping, throw- ing up all the time.

Jennie stopped at the fifth tick because she thought she had arrived at the external plot point that would kick off Ruby's story. And she *had*—it's true. But do you see what's missing in the fifth tick? *Something to force Ruby to act.* I asked Jennie to go one tick beyond this—to the tick that brings us to the dog. Here's Jennie's sixth tick.

Sixth Tick: Because Ruby's so distraught, her sister Nora—the veterinarian, who believes Ruby should have had a dog all along, if not a husband—is worried that she's losing it. Nora is, in fact, convinced that Ruby is a danger to herself. And so she has decided she's going to save Ruby from herself by removing her from her house and sweeping her off to wherever Nora lives. (I know, I know, people will wonder how in the world one adult would have the power to do that to another. I'll figure something out. . . .) In any case, in Ruby's mind this is the final straw. This is an indignity and an inconvenience and an insult she simply will not tolerate—which is precisely why she will, in the *next* scene, decide it is a good idea to "borrow" a dog, just for an hour or two, to convince Nora to leave her alone so she'll have a chance in hell of doing this impossible thing she has to do. Instead, it will force her to flee with the dog. Needless to say, in her grief-crazed state, Ruby's decision-making abilities are clearly a tad off-kilter.

Aha! The sixth tick has it. This tick drops the reader straight into the external plot problem (the script must be revised) and straight onto the third rail (at this point Ruby doesn't trust *any* kind of connection—to family, to fans, to dogs). There's no preamble, no explanation, just the thing itself. But as you'll see when Jennie writes this scene in the next chapter, we'll get a lot of what's in those first few ticks as Ruby makes sense of the situation she finds herself in—all of it in service of what is happening right now.

And plotwise, since Jennie already knows that the coming dognapping scenario has a pretty potent ticking clock of its own, which will quickly send Ruby on the run, she can now harness it to Ruby's overarching problem and yell "Mush!"

It is a perfect example of the protagonist believing she's about to sail into a handful of the ice cubes we were talking about earlier, rather than slam into a massive iceberg. Because as far as Ruby's concerned, she'll snatch the pooch for an hour or two, just long enough to convince Nora to leave her alone, and then return it to its backyard, no harm done. It won't work out that way—which, of course, is the point.

It is also a perfect example of how the surface events are not what the story is about. On the surface, dognapping never had anything to do with Ruby's story. But then, what did broken plumbing have to do with the protagonist overcoming her agoraphobia in *Sparrows Dance?* In and of itself, not a thing. The filmmaker used it to compel his protagonist to come to grips with the problem that plagued her by forcing her to come into contact with another human being, just as Jennie will use Ruby's flight with the dog to force her to deal with her misbelief by bringing her into contact with people who challenge, and ultimately change, her worldview. And at the end of the day chances are *that* is what will allow Ruby to then write the script and resolve both the plot and the internal story.

WHAT TO DO

Now it's your turn to sketch out the ticks that will lead you to your opening scene. Your goal is to find the tick that catapults your protagonist into unavoidable action. You'll know it when you get there, because you'll feel a strong tug of forward momentum—a sense that your protagonist *must* act and must act *now*. Keep the clock ticking until you get there. Don't be afraid to try this again and again until you get a tick that has everything it needs—the overarching plot problem, the main ticking clock, the third rail.

Once you've nailed that tick (could be your third or fourth or seventh tick—there's no prescribed number), you'll be ready to delve into the opening scene of your novel, which means you'll now be blueprinting in earnest.

Before we move on, however, take a second to congratulate yourself. By diving into your protagonist's past, you not only know when your novel starts, but you know why, you know what's at stake, and you know what it means to your protagonist. Even more amazing, you know what your novel is about. That, my friends, is surprisingly rare. Usually, when you ask a writer what their novel is about, they eagerly give you a rousing rendition of what happens in the plot. But as earnest and passionate as they are about their project, it soon begins to sound like a bunch of things that happen, instead of a story. You listen, first trying to keep it straight in your head, and then simply to keep your mind from wandering, but you can't help thinking, why would any of this matter?

You, on the other hand, have the makings of a riveting novel, because you already know a good bit about the story that it will tell. Now, the goal is to begin creating a plot that will bring it to life.

One last thing: From here on out we're going to focus much more on writing than on brain science. Because now you understand what hooks readers, and you know *why*. You know how the brain works, and you understand the unparalleled power of story. You know that story is the language of the brain. From here on out, you're not going to be studying that language, you're going to be speaking it.

PART III

CREATING AN EXTERNAL GAUNTLET TO SPUR YOUR PROTAGONIST'S INTERNAL STRUGGLE

9

THE OPENING: OF YOUR NOVEL AND OF THE STORY GENIUS BLUEPRINTING SYSTEM

Never confuse movement with action.
—ERNEST HEMINGWAY

So what exactly is a novel blueprint, and how does it help you create a plot that touches on the third rail? I thought you'd never ask.

In a nutshell: A novel blueprint is a scene-by-scene progression of your external plot, as driven by the internal struggle each event triggers in your protagonist.

You'll begin to build your blueprint based on everything you've unearthed in the process thus far. You've already seen how your protagonist's past has guided his life up to the tick that will open your novel. Now it's time to plumb that past, often digging yet deeper into it, the better to nail down the events it will thrust your protagonist into, along with his reaction to them.

If you think harnessing the plot to the third rail will be tricky, here's a welcome news flash: you've already been doing it for quite a while now. Take a look at the progression of backstory scenes you've already written. In each scene, what happens externally is driven by, and gets its meaning from, your protagonist's internal reaction to it. For instance, Jennie's origin scene doesn't simply recount external events—Mr. Anderson's death and Beth's breakdown on the soccer field—it recounts, first and foremost, how those events affect Ruby. The scene is about the inception of her misbelief, the misbelief that will guide the external trajectory of her life: that love is what ultimately destroys us. The same is true of your protagonist. If you're writing a mystery, this will apply to your bad guy, and perhaps your investigator as well.

Here's the secret: although your blueprint (not to mention your novel itself) will be made up of individual scenes, in truth those scenes are not individual at all, but part of this escalating cause-and-effect trajectory. Each scene will be triggered by the one that came before it and will trigger the one that follows. That's why even though you'll work on each scene and each plot point separately, you must always be keenly aware of the part it plays in the overall trajectory. Otherwise even the most brilliantly executed scene will not only stop your story in its tracks but also be incapable of doing what it must do: make the scene that follows it inevitable. The blueprint you're going to construct will make that mistake impossible, and it will make writing each scene in your novel far more intuitive than you ever imagined.

But since each scene is linked to those that precede and follow it in a single seamless, escalating storyline, does that mean you need to lay it out in chronological order on a massive roll of butcher paper stretching across the living room into the bedroom and then under the kitchen table as you scribble forward? Hardly. That's cumbersome, and so last century (do they even have butcher paper anymore?), and most importantly, it's not flexible enough to allow you the leeway you need for the development process. Instead, we're going to use a new twist on an old method: Story Genius Scene Cards.

I'm betting you've heard of index card–type models before. There are many versions out there, all of which imply—or state outright—that the cards are written at random with no prep at all, and can be, *should* be, moved around at will. In fact, that's usually a major selling point. Take your cards! Throw them in the air! Pin them up in new and different ways!

That method is deeply flawed.

When it comes to cause and effect, if you move one scene, chances are the others will cease to make sense, and the internal logic of your novel will be tangled into an unfathomable knot. And yet, at this stage you *will* be moving scenes, and—far more frequently—adding scenes, so how do you do that without destroying your novel's narrative chain?

It's simple. By tracking how each event is linked together in an escalating, causal succession. That way you'll know at a glance which events you can easily move, and which ones, should you move them, would cause your plot to collapse on itself like a house of cards. This also will alert you to any logical gaps and internal inconsistencies left in the wake of a move—think of it as literary triage—thus allowing you to remedy them, ere they become the sinkhole that swallows your novel whole.

You might be worried that a plot built in this way will become too rote, too predictable, and that it might bore you, let alone the reader. Fear not. For one thing, there are almost always several possible logical outcomes for every scene. Knowing what those possibilities are is what spurs the reader forward, anticipating what they hope will happen, fearing what they hope won't.

And while we're calling them cards, let's be honest. As much as writers may love the thought of quill pens, thick creamy paper, and ink-stained hands, the strictly analog world has its drawbacks. Specifically: An index card is 3 by 5, which means unless you write with a magnifying glass, and your vision is so acute you really can see all those angels boogying away on the head of a pin, it's just not big enough. Plus, ink is pretty much permanent, and you can only erase pencil so many times before you've shredded the damn card.

Thus you may want to go digital, and create a template on which you go as deep and get as specific as you like. If you prefer to stay in the analog world, better to use a sheet of paper than an actual index card. And now, without any further ado, here's the Scene Card template you're going to use from here on out:

SCENE #		
ALPHA POINT:		
_____ SUBPLOT:		
_____ SUBPLOT:		

	CAUSE	EFFECT
	What happens	**The consequence**
THE PLOT		
	Why it matters	**The realization**
THE THIRD RAIL		
		And so?

The first thing you'll fill in is what I call the scene's Alpha Point. It refers to the key role the scene will play in your novel's external cause-and-effect trajectory. It's related to action, to the seminal thing that happens. The question the Alpha Point must answer is, why is this scene necessary? What is its main job? This is what anchors the scene in your novel's timeline. Every scene must have a concrete Alpha Point, and when you start a card for a scene this will often be the *only* blank you can fill in. That's perfectly fine. However, scenes do much more than make a single point, or cause a single change. Most scenes will move several subplots forward at once and cause numerous changes that will ripple through your novel. So beneath the Alpha Point is space for multiple subplots. As you build your blueprint, adding layers and developing subplots, you'll continually be adding info to existing cards. But the Alpha Point remains the most crucial point, because without it the scene is just a random thing that happens.

Notice the vertical line that divides the card in two.

- The left side represents the *cause* side of the cause-and-effect equation.
  - For the plot, that's what happens in the first half of the scene.
  - For the third rail, that's why what is happening matters to your protagonist, given his or her agenda.
- The right side is the *effect* side.
  - For the plot, that is the external consequence of what happens in the scene. Be very clear: this is the consequence that takes place within the scene itself, *not* the consequence it will have in the next scene.
  - For the third rail, it's the internal change, the realization that the event triggers in the protagonist. (Or, if the protagonist is not in the scene, the realization it triggers in the scene's POV character, *and also* the realization it will trigger in the protagonist when he or she finds out.) This realization must lead to action—often to the very next thing the protagonist does. In other words, the realization causes him—in ways large or small—to shift his game plan.

Keep in mind that *every* scene must produce a hard-won change, both externally and internally. If you're writing a mystery, courtroom drama, or police procedural the same is true: every turn, even if it's realizing that a promising lead fizzled, triggers an internal insight that helps the protagonist better understand what's going on.

Naturally, the plot-level change is easier to see because it is, well, external. That is, you can actually *see* it—kick a can, it skitters down the street; rob a bank, the cops will chase you; forget to feed your pet hamster, and, don't ask. So it's easy to miss the crucial fact that your protagonist's *worldview* must also change a little bit in each scene as he or she struggles with what to do, what action to take. Every scene involves an internal struggle that touches your novel's third rail. These are not "separate" struggles, but turns in the wheel of a single, escalating struggle from the first page to the last. Noting how your protagonist changes a little in every scene not only helps you keep track of their evolving worldview, it also gives you insight into what they'll do next, and of course, the most crucial element: why they'll do it.

If your novel is in the third person, there will be scenes in which your protagonist is offstage and doesn't appear at all. In that case, everything in the upper portion of the card—the plot—remains the same. But in the lower portion—the third rail—you'll answer the question on the left, *Why does it matter?*, for the scene's POV character *and* for your protagonist. On the right side, where you enter the realization, you'll do the same—note the POV character's realization and note what the protagonist will realize *when she finds out what happened.* Sure, your protagonist might not be aware of what happened in the scene for quite a while, but you need to know how it will impact her when she finds out. After all, that's exactly what your readers will do as they weigh the impact the event will have on your protagonist's struggle.

On the very bottom of the card is the space for answering, *And so?* This is where you'll note what must happen next, as a result of what occurred in the scene. Like the Alpha Point, it needs to be a concrete, action-based event, rather than something conceptual and emotion based. For instance, if the consequence of what happens in the scene is that Ruby learns that Henry is a goner, the *And so?* wouldn't simply be *Ruby is very, very sad.* The *And so?* needs to answer the question, what is Ruby going to *do* as a result of being sad, especially since losing Henry means she not only is bereft but also has to revise the script alone? Thus the *And so?* helps you lock each scene into place, allowing you to monitor your cause-and-effect trajectory, so your novel remains true to your story's internal logic.

The completed Scene #1 card on the following page shows how Jennie worked up the opening scene we watched her develop in the ticking clock exercise, so it's already very clear and defined.

Why This Scene Card Works:

- Although a lot happens in this scene, Jennie's Alpha Point captures and concretizes the scene's main point. Ruby is going to take action and write the script (or so she thinks). That is the point of this scene, the thing that needs to happen or the story can't go forward. Because hey, otherwise for all we know Ruby might have sat in her house, sobbing, forever, and who wants to watch that?

- The subplots are clear and concise and in a few words pinpoint Sharon's and Nora's roles in the scene, and what each learns from it, allowing Jennie to anticipate what they might do next. There may be times when there is no subplot to record, and that's fine too.

- The "What happens" reflects the initial action in the scene, the initial thing your protagonist must respond to. Sometimes this column will have only one event, sometimes several. Here, that initial action is the call Ruby receives from Sharon, followed by a visit from Nora.

- "The consequence" within the scene is a direct result of "What happens." Ruby changes before our very eyes: she is going to write the script, and to do that, she first has to placate Nora.

- This change isn't just surface action; it's driven by "Why it matters" to Ruby: she is bereft, lost, at the end of her rope . . . but there is *one* thing that she still has control over that matters to her: having the

SCENE #1

ALPHA POINT: Ruby decides she must pull herself together and rewrite the script

NORA SUBPLOT: Nora realizes Ruby is a risk to herself

SHARON SUBPLOT: Sharon realizes Ruby's in denial and so a fangirl might have to write the ending

	CAUSE	EFFECT
	What happens	**The consequence**
THE PLOT	• Sharon tells Ruby that if she can't rewrite the show's ending on time— the deadline is in 72 hours— then a fangirl named Clementine will, and Ruby will lose the one thing she still has control over. • Ruby's sister, Nora, shows up unexpectedly and expresses mounting concern for her well-being.	• Ruby promises Sharon she will rewrite the finale. What she doesn't say is that she has no idea what it will be, and she hasn't written a word since Henry's accident. • Nora's concern feels like an intrusion, rather than an act of caring. When Ruby tries to blow her off, Nora announces she's coming back tomorrow to take Ruby to her house.
	Why it matters	**The realization**
THE THIRD RAIL	• Ruby's been distraught since Henry's accident; she can't eat, sleep, work; she's about to give up. • Ruby has never written a script without Henry. • Ruby has no experience with this kind of pain, especially since she never thought she'd feel it. • The show is the only connection she still has to Henry, and to the work that's sustained her.	• Ruby realizes that Nora means business, and is much more of a threat than she'd thought. She decides to find a way to convince Nora that she may be sad, but otherwise she's perfectly fine. **And so?** Believing that clean clothes and a full refrigerator will convince Nora she's okay, Ruby heads to the market.

final word on the show she and Henry created. That is something she will not give up.

- Ruby's "realization"—the thing that will drive the *next* action she will take—doesn't have to do with the show, per se. It has to do with the only thing she *thinks* is standing in her way: Nora's concern.
- The *And so?* pinpoints exactly what Ruby's next action will be, and why she thinks it's a good idea.

WHAT TO DO

Now it's your turn to create a card for your opening scene. Remember, your goal is just as much to be specific about your protagonist's inner struggle as it is to be specific about what will happen in the scene. The two are linked, and each is neutral without the other. Your protagonist's internal agenda is not simply what gives emotional weight and meaning to what's happening up there on the surface; it's also what drives the decisions she makes, and therefore the action.

The Essential Process: Going from a Card to Writing a Scene

You may have found that filling out a card is not a simple, static exercise. It's going to crack open your story each time. Things might get messy, because concretizing the internal and external trajectory of each scene often reveals the gaps in your story logic—and what you still need to dig for.

Here's Jennie talking about everything she thought about after filling out her first card.

> &*$^#, I thought I had this scene so dialed in, but now I see that I can't write it until I know things about Nora (where she lives, why exactly she's showing up on this day, why she has this power over Ruby); and I can't write it until I know exactly what happened to poor Henry (Is he in

a coma? Brain dead? Facing multiple risky surgeries?); and I can't write it until I know who else would be in the hospital with Henry (Friend? Brother? Mother?); and worst of all, I can't write it until I know a lot about their TV series, the current season, and how Ruby and Henry originally ended the final episode, which means I have to design a whole flippin' TV show *and* figure out their entire writing history. I think I'm going to cry.

This might be a good time to remember that writing a novel is hard. You are creating an entire world where nothing existed before. Each decision is going to lead to a whole slew of questions. Every "what happens" will send you digging for a "why it happens." This process of developing Scene Cards, raising questions and answering them so that you can write the scene itself, is the essential component of blueprinting and what will turn you into a Story Genius.

Right now, I'd like you to see what it looks like to go from sketch to card to fully fleshed-out scene, so for the moment we're going to skip over the work Jennie did to answer those questions about Nora, Henry, and who's standing guard over him at the hospital, not to mention the story behind Ruby and Henry's TV show—we'll come back to it all in later chapters, as we explore how to anchor your scenes in the specifics that give them meaning.

When you begin writing forward and you know there's something you have to figure out, but *not now*, simply put TK—a proofreader's mark that stands for To Come (and is easier to spot than TC)—and keep on going. You can describe your TK—TK Year, TK Destination—or insert TK all by itself. Once you've finished the scene, you can go back to those TKs and begin digging for the specifics they're now standing guard over.

With that in mind, here is an early draft of Jennie's first scene. It's still a work in progress, but it's got all the bones it needs to support itself, and give rise to the story to come.

Write or die. That's really what it came down to. Because how could I live with myself if I didn't finish the show? How could I face Henry and the doctors and his mother, who sat hour after hour by his bedside listening to the machines beep and wail, watching his chest rise and fall, if I didn't write one last script?

He was there because of me. Because we fought and he got on his bike and I let him go up into the hills where some kid with twice the blood alcohol limit going twice the speed limit looked the other way. Or didn't look at all. Or looked and decided to plow right into the man with whom I've spent the last seven years making a life word by word by word.

"No divine intervention," Henry always said when something in a script seemed to come without being earned.

So what the fuck was this? In the midst of filming the final episode of a seven-season run? When we had finally decided to write a twisted version of *Romeo and Juliet*? It felt like divine intervention of the worst kind. It seemed too cruel to be true.

I got up and went to the garage and stood there in the slightly cool air looking at the red rubberized hook where his bike had always hung. He had his mail delivered to his house, but he kept his Cannondale here. I stared at the vacant red hook as if I could conjure up the bike, as if I could reel back time, as if I were the god in charge of the script of my own life.

In the revision, he wouldn't get on the bike. We wouldn't have had the fight. I wouldn't have been so bitter about his decision to end our golden run.

Before the start of the season, Henry had been invited to write a big-budget movie script—just him. Not me. I took it to mean that they thought he was the real writer, and that I was just there for the PR. We had always been the kind of feel-good story Hollywood loved—two former behind-the-scenes crew members stepping out to become the scene makers. But I imagined everyone whispering, *She just polishes things up. He does the heavy lifting.* The producers of the movie probably thought they could get all the writing magic for half the price. And instead of refusing the job, Henry accepted and then proposed we do *Romeo and Juliet* as our swan song—the play we'd always assiduously avoided because it terrified me. Because it cut so close to the bone. I'd said yes, because I felt so gut punched, so bereft, so lost at the prospect of not building my stories—and my days—around him. At the time, that reality—not writing with him—felt like the worst possible thing I could imagine.

I shut the garage and went to the refrigerator, where there was nothing but a little plastic container of applesauce. I got it out and forced myself to eat a few bites. Write or die. If I was choosing to write, I had to eat.

Just then the phone rang.

Let it be TK Name, I thought. Henry's TK Relative. Let TK Name be calling to say that Henry was awake. He was fine. He was asking for me and I should come right away.

I grabbed the handset.

"Hello?" I said—all my hope and fear crashing together into a sound like an explosion, like a demand.

It was Sharon, my producer. She'd been calling three times a day, asking in a too-kind voice how it was going, asking if I needed anything. I had revised three pages of a forty-eight-page script that was due into the cast's hands in seventy-two hours, and it was getting harder and harder to lie to her.

"How's it going?" Sharon asked.

"Fine," I said. "Great."

"Tell me the truth, Ruby," she said.

I gazed across the room at my laptop open on the big wooden table. Henry and I had written a script where Juliet gets pregnant, and Romeo orders the poison because he can't stand the thought of being as boorish a father as his own dad. She talks him out of killing himself, but then she dies in childbirth, and Romeo is left to face his biggest fear and his biggest loss every day of his life. I had never liked it. I thought it was sad without being tragic, and there was no satisfaction in that.

"You're just being contrary because you're mad at me for taking the Siegel project," Henry had said.

"I am mad. It's crap that they didn't ask me, too. Just because you do all the social media, people forget about me. They think I'm—what? Your typist? Isn't that what they say?"

"That's just ignorant fans speculating," he said. "Why let that get to you?"

"Because apparently the movie producers think the same thing," I said, and looked away.

"Ruby," he said, "I told you it wasn't my place to bring you on board. They asked for me. Just me. I don't know why, but what was I going to say? 'No, unless you hire Ruby, too?'"

"That's an idea," I said.

He stood and took my hands in his. He lifted them and kissed them. He took me in his arms. "Ruby," he said. "It won't change anything between us."

I laughed, because it would change everything.

We made an uneasy peace while the show went into production, but it all came to a head two weeks before the end of filming. The cast and crew were assembled, everyone was being paid for their time—and Sharon's boss suddenly decided he didn't like the ending. He, too, thought it wasn't quite edgy enough. He ordered a rewrite—while everyone waited for our words, while the clock kept ticking.

"Let's let Juliet live," Henry proposed. "Let's let them live happily ever after."

"No," I said. "No way."

"Don't let your own feelings get in the way of the story," he said—and I lost it. I yelled and screamed at him, told him I didn't need him to tell me how to write a good story, told him I didn't need him at all.

"This is bullshit," he said, and he put on his riding clothes and got his bike down from the red hook and pedaled out from my house that needed painting and my garden that needed weeding, and up into the hot hills where that kid was driving like he believed that nothing in the world could ever hurt him.

"Ruby?" Sharon said into the phone, "You still there?"

She had asked me to tell the truth about the script. "The truth?" I said, "The truth is that I'll get it done."

She exhaled so loudly I could almost feel it. "Listen," she said, "There's no easy way to tell you this. But I met with Jim and we need a firm commitment from you by the end of business today. We have a Plan B in place, and if we need it, we need to act on it right away. Of course that's not what I want. It's not what any of us wants." She paused, then said quietly, "But if you can't deliver the revision, Ruby, someone else is going to have to."

So there it was—the ultimatum. I dropped the phone and vomited my applesauce into the sink.

"Ruby? Ruby?" I heard Sharon calling through the phone—the cry of someone who couldn't do a damn thing to help.

I spit into the sink, then picked up the phone. "Sorry," I said.

"Oh Ruby," she said. "Can I send a doctor over?"

I laughed. Was there a procedure to leech guilt from your veins? To help you finish a story you couldn't bear to end?

"I'm worried about you," Sharon said.

She sounded exactly like my sister, Nora—*Are you eating, are you sleeping, have you gone outside today?* They both sounded as if they were talking to a sick child or a wounded animal, and what good did that do? These were not soft women. Nora raised horses and dogs and children and ruled them all with her no-nonsense logic. Sharon ran a rogue TV empire that made long-time studio execs weep. Toughness was an attribute we all shared, and I was disappointed that they would revert so quickly to pity.

"Don't worry," I said. "I'll meet the deadline." I took a sip of water to rinse out my mouth. "But just out of curiosity, what's your Plan B?"

"Clementine," she said.

"Clementine?" I shouted. The most die-hard of all the fangirls. Clementine, who started the most popular of all the *Complications Ensue*

fan sites, where she analyzed every show, held live debates and Twitter events, and invited people to share their own versions of twists we hadn't offered, their own endings, their own scripts. She was often quoted by the press in articles about the show, and she would talk as if she were an authority, as if she had some kind of inside track just because she exchanged little tweets and Facebook posts with Henry. Of all the people on the planet, she was the last one I would want to take over my show.

"You can't be serious, Sharon."

"We're bringing her into the studio tomorrow to work on the script. That way we'll have something by Sunday, just in case the rewrite is too much for you."

"It's my show, Sharon," I spit. "Mine and Henry's. I'm finishing it if it's the last thing I do. You can bring Clementine in and polish her script all day, but my script is the one you're shooting on Monday. Is that a clear enough commitment for you?"

"Clear as day," she said, in a voice that made it seem like she didn't believe for a second I would pull through. "Let me know if you need anything."

I hung up and slid to the floor. I wanted to call Henry to tell him to tweet Clementine and tell her to back off. I wanted to tell him not to worry—I would rewrite the ending.

I rested my head-on my knees and suddenly there was a knock on the door. "Ruby?"

It was Nora. I dragged myself to the door and eased it open.

"I had to come into town to get dog food," she said, and stepped inside. She was wearing dirty jeans, cowboy boots, and a flannel shirt rolled up at the elbows. When she took a step toward me, I saw a chunk of what looked like horse shit fall off the heel of one boot onto my wooden floor, and I flinched—not because I cared about the state of my floors, but because it was clear that Nora had come straight from the stables, as if she thought the house was on fire, as if she thought I needed to be rescued. The dog food errand was a total lie. She was, after all, a vet who probably got specialty food delivered in bulk.

I watched as she went to the refrigerator, telling me she'd brought a green smoothie, some quinoa, some soup—and in that instant, I realized that in the hierarchy of things that were threatening to do me in, Nora posed a much greater threat than I'd thought.

She went to the bathroom, came back, stood there a moment, and then cleared her throat. "Sleeping pills?" she asked.

"What are you talking about?"

She pointed to the bathroom.

"So what?" I said. "I'm having trouble sleeping. Sleeping pills help with that."

"You've never taken sleeping pills before, have you?"

I laughed—a rough little laugh. "I've never had Henry comatose in the hospital with seventy-two hours to rewrite a show ending, either."

Nora took a deep breath. "Mike and I talked about it, and we're really worried about you. We don't think you should stay here alone. We want you to come out to Ojai and stay with us—just for a few months, just until . . ."

I forced myself to take a breath. "It's sweet of you to be so concerned," I said, "but I'm fine, everything's fine. I'll eat the food you brought, I promise."

"I really don't think you should stay here by yourself, Ruby. It's not good for you right now," she said in a soft, caring voice that worried me even more. "I'm coming back tomorrow, okay? I'll help you pack up some of your things."

I said nothing as she gathered up her things and left, but there was no way I was leaving the house and no way I was letting some fangirl steal our show in the final stretch. I was going to do the rewrite, and I was going to do it here, at the big wooden table in front of the bay window where I always wrote, while Henry sat and stood and paced, and cajoled and questioned and pushed, and refined my ideas so that instead of being vague and thin, they became something whole and alive.

All I had to do was convince Nora to leave me alone.

Why This Scene Works:

- Did you notice that in the very first sentence Jennie told us what was at stake? That one three-word sentence, "Write or die," instantly gives us a context that will give meaning to everything that happens from that moment on. It also instantly engenders the reader's curiosity: *Write* what? *Or die? Something big must be going on here; I wonder what it is . . .*

- Did you notice how much of this scene is backstory? That is, Ruby's memories—the yardstick she uses to make sense of the present, as she struggles to make a difficult decision? Point being: All the information Jennie uncovered in the first several ticks of the clock (and way before that) is an active part of the lens through which Ruby evaluates what's happening now. And, most importantly, it is based on that subjective, strategic evaluation that Ruby then takes action.

- Did you notice that Jennie used Ruby's notion of how to rewrite R&J to give us information about her misbelief about love?

- Did you notice that Jennie was very clear about Ruby's immediate agenda? To wit: Ruby realizes that—as far as she can see—the biggest impediment to her turning in the script on time is her sister, Nora. We readers, of course, are way ahead of Ruby—which is exactly where Jennie wants us to be. Because she knows it will allow us to anticipate what might happen next, and engender empathy for poor, misguided Ruby, who's clearly in for a rude awakening. In this case, we instantly know two things: (1) Nora is most definitely *not even close* to the biggest problem Ruby has; and (2) given Ruby's state of mind, whatever plan she comes up with is pretty much guaranteed to make the situation worse.

- Did you notice that as a result of Jennie's showing her hand so openly, the first scene immediately drives toward the next? That is, Ruby is going to come up with a plan to deflect Nora's unwanted attention. That, my friends, is *both* layers of the novel's cause-and-effect trajectory at work. We know *what* Ruby is going to try to do next, and we know *why*.

- Did you notice that although there is no dog in this scene—not even a hint of a dog scratching at the door—Jennie is planting the seeds that will support it? How? Note the way that Ruby talks about Nora (she's a veterinarian, she raises dogs), facts that will give a warped kind of credence to Ruby's coming brainstorm: *If I have a dog, it'll convince Nora that I have an emotional support system in place—that I haven't lost my mind!*

WHAT TO DO

Now it's time for you to write your opening scene. Don't worry. It's not about "getting it right." Because at this stage, you can't, so try to relax (as much as is possible when staring at that damn blank screen). What you're going to write is the first layer of a scene that will probably be rewritten more than any other in your novel. Why? Because it's where the seeds of what will happen throughout the novel will be planted. You don't yet know what all those seeds are, so by definition you can't plant them all now. But you can create the field, till the soil, and get it ready. Just have your stash of TKs ready to help you pinpoint exactly where you still need to dig—and what info you still need to dig for.

10

THE REAL "AHA!" MOMENT: WHERE WILL YOUR STORY END?

If I see an ending, I can work backward.
—ARTHUR MILLER

The next question we're going to tackle is, where does your story end? *End?* you may be thinking. *Hey, we only just got started.* True, but unless you know where your story is headed right out of the starting gate, chances are you'll never get there. Which isn't to say that the end you come up with now will actually *be* the end. But if not, it will be because somewhere down the line, you had a specific story reason to strike out in another direction.

After all, we can never really be sure of the future, can we? The future is where our goals lie, shimmering in the distance, daring us to take chances, the better to achieve them. Sometimes that goal is the bright shiny thing we're galloping after; sometimes it's the flood we're trying to outrun. Maddeningly, the specific road that leads us to our goals (or, tragically, in the other direction) is clear only in hindsight. As far as the reader is concerned, the same is true of your protagonist, but—and this is the point—you, the author, need to know what the future holds for your protagonist right now *in order to create the road to get her there.*

You already know what your protagonist's long-standing goal is, and you know the unavoidable external problem that life is about to toss into her lap, forcing her to leap up and hit that road. What's more, you've hooked it up to a nifty ticking clock, so whether it's set to track years or a single day, time is already running out for her. Now the question is, what will *actually* happen when she gets to the end of that road?

The answer is worth its weight in crumbling erasers, worn-down Delete keys, and months, if not years, of rewriting. It will guide you through the creation of your plot, immediately helping you pinpoint and shape the escalating obstacles that your protagonist will have to scale, obliterate, or cleverly reframe in order to arrive at the conclusion that you're aiming for, the point that you're making. Lest you think this is just a move for newbie writers, here's what veteran writer and National Book Award winner Joyce Carol Oates has to say about it: "I always know what the end is, and where I'm going; I'd never just sit down and start writing. That would be like getting in a car and just driving, with no idea where you're going—no serious writer would write like that."[1]

That's why in this chapter we'll learn the secret of what "the end" *really* means (hint: it's not about the plot), how the third rail is transformed by in the protagonist's "aha!" moment, and how the "aha!" moment is earned in reaction to what the plot has put your protagonist through.

Why Write the End So Early in the Process?

Like the opening scene you just wrote, the scene you'll write in this chapter will simply establish the first layer of a scene that ultimately will be rewritten many times over. The funny thing is, this initial depiction of the final scene is the most crucial version you'll write, because it will help you begin to figure out what needs to happen between the first page and the last page to make sure that your protagonist works overtime to earn her "aha!" moment. Since even at this early stage you've already probed deeper than many writers *ever* go, you have enough inside info to write a scene that won't be general, but will instead be full of specifics that will help you identify exactly what to develop as you build your blueprint. This scene will guide your story home—from the very first moment of the very first scene.

In fact, often the very first scene in a novel includes a glimpse of what the ending will be. Contrary to the popular (and erroneous) notion that this "gives it all away," letting readers know where the novel is headed is actually the very thing that lures them in. And it goes without saying (she said anyway) that the only way to do that is to know the ending up front, allowing you to take John Irving's advice: "Whenever possible, tell the entire story of the novel in the first sentence."[2] For instance, here's the opening sentence of Irving's own *A Prayer for Owen Meany:*

"I am doomed to remember a boy with a wrecked voice—not because of his voice, or because he was the smallest person I ever knew, or even because he was the instrument of my mother's death, but because he is the reason I believe in God; I am a Christian because of Owen Meany."[3]

Sage writers often open with a clear vision of where their story is headed. For instance, the second paragraph of A. S. A. Harrison's gripping psychological thriller *The Silent Wife* neatly sums up what Jodi, the protagonist, is in for:

At forty-five, Jodi still sees herself as a young woman. She does not have her eye on the future but lives very much in the moment, keeping her focus on the everyday. She assumes, without having thought about it, that things will go on indefinitely in their imperfect yet entirely acceptable way. In other words, she is deeply unaware that her life is now peaking, that her youthful resilience—which her twenty-year marriage to Todd Gilbert has been slowly eroding—is approaching a final stage of disintegration, that her notions about who she is and how she ought to conduct herself are far less stable than she supposes, given that a few short months are all it will take to make a killer out of her.[4]

Or how about Donna Tartt's deeply, deliciously disturbing literary debut novel, *The Secret History,* which opens thusly: "Does such a thing as 'the fatal flaw,' that showy dark crack running down the middle of a life, exist outside literature? I used to think it didn't. Now I think it does. And I think mine is this: a morbid longing for the picturesque at all costs."[5]

That sentence not only sums up what the novel is going to be about, but identifies the protagonist's misbelief. In fact, Tartt gives us a glimpse of the lengths to which that "fatal flaw" will drive him *before* chapter 1, in the form of a two-page prologue in which it's pretty clear that the narrator, Richard, is going take part in the upcoming and, yes, morbidly picturesque murder of his erstwhile chum, a fellow affectionately called Bunny.

All three of these novels give the reader the big picture right off the bat, supplying them with an overarching context that gives direction and meaning to what follows, deftly arousing our curiosity, inducing that intoxicating rush of dopamine. Remember, the first job of a story is to make the reader *want* to know what happens next, and what lures them in is generously allowing them to anticipate just what "next" might be.

But right now, for you and Jennie, there *is* no firmly established next. And here's something interesting (not to mention heartbreaking): writers very often stop writing after the first twenty pages because *they* have no idea what comes next, either. The problem is that *because* there are so many options, it's the same as having none. It soon begins to feel so overwhelming that they decide to take a little break from writing, you know, for a day or two so they can regroup. And we all know how *that* one ends.

That's why we're going to dig into what your ending is really about, so you know exactly what you're aiming for as you write.

What "The End" Is Really About

You might be thinking, *Wait a minute, haven't we done this already? Didn't we do that when we worked out the protagonist's misbelief and what she is struggling to achieve? Isn't that what the ticking clock is leading to?* Absolutely. Your plot will revolve around what your protagonist will do, starting on page one, in reaction to that rapidly approaching deadline.

But guess what? The ending we're looking for here (and what your reader is most hungry for) isn't just about what happens plotwise—it's about what your protagonist *realizes* as she faces it head-on. The problem is that writers, even experienced writers, tend to fall into a very seductive trap: envisioning the end of the novel simply as what happens externally, rather than what the protagonist learns.

It's an easy mistake to make. After all, when you think of the ending of a novel or a movie, what comes to mind? I'm guessing it's the very last scene, when the external plot problem is finally resolved, and the protagonist can at last savor, or rue, the new life he or she is about to embark on. It's Benjamin Braddock breathlessly leaping onto a rundown city bus with Elaine Robinson in an elaborate wedding gown, having just ditched her would-be groom to run off with him in *The Graduate*. It's Rosemary ditching the really rotten devil she knows, to rock the cradle of the teeny tiny devil she doesn't, in *Rosemary's Baby*. It's Rick lovingly ditching Ilsa, leaving her with one last kiss (and a very quotable speech) before putting her on a plane and then wandering into the fog with Renault, at the start of a beautiful friendship, in *Casablanca*.

Each of these scenes is incredibly moving, very visual—and not what we're talking about. Because they are about how the *plot* wraps up. And the plot, by itself, is just a string of external events. Which is why those endings wouldn't amount to a hill of beans if we didn't know how they were affecting the protagonist. They move us as readers or viewers only because we know exactly what they mean to Benjamin, Rosemary, and Rick. We know what it cost each one of them to get there. We know how the journey from the first scene to the last changed their view of the world and of themselves, and therefore how they feel about what's happening now, as the story ends.

So even though your protagonist's "aha!" moment might indeed occur just as the external problem is solved, that's not what the scene is *about*. It's actually about what the event has taught your protagonist. The moment you want to capture on paper is when your protagonist's internal struggle ends, as her misbelief finally bites the dust and she sees the world with new eyes—aka her "aha!" moment. Often, it's what allows her to finally solve the external problem, or make peace with it. And as with most things, there's more to it than meets the eye.

Getting the Revelation Right

It was life experience that convinced your protagonist to staunchly embrace her misbelief, and throughout the novel current life experience will try, with escalating force, to disabuse her of that notion. Now, she's at the place where life is going to clobber her with its best shot. She's going to either finally see things as they really are and so come out battered but victorious, or, alas, go down for the count.

This is the scene in which she either makes that internal change or doesn't, for as long as the novel shall live. Amen.

Here's the secret: The point is not *that* she makes the change, it's *how* she gets there—internally—that counts. And ironically, even when writers *do* get everything else right, it's the logic behind the internal change that often goes missing. There are three ways to make sure that doesn't happen to you.

Let the Protagonist Earn the Revelation

The deep satisfaction readers feel as a novel ends is based not on what the protagonist has achieved externally, but on how he's changed internally, giving him the insight to make it happen. In other words, it's not that he solved the problem, it's what he learned in the process. That's what your readers have been tracking from the get-go, and this is where they viscerally experience the epiphany you've been building toward from page one. That's why you can't impose this final, life-altering revelation on your protagonist, but must allow him to earn it himself, right there on the page.

So that you don't inadvertently step over it, it helps to keep in mind that while the "aha!" moment always comes late in the novel, it doesn't necessarily come at the very end, when the plot itself concludes. Sometimes it comes just before the end and is what gives the protagonist the courage to face that final, often excruciating external hurdle. Sometimes it comes at the very moment the protagonist is locked in that last all-out battle, and it's what gives him the courage, strength, and wisdom to keep going against all odds. And sometimes the "aha!" moment comes right afterward, as the protagonist is making sense of what just happened, surprised that it leaves him feeling very differently from what he'd expected.

Put the Reader in the Midst of the Event Itself

Your goal as a storyteller isn't to tell us what your protagonist realizes; it's to plunk us into the event that causes her to have the realization in the first place. A mistake writers often make is leaping into the scene too late. So they begin just as the protagonist has *had* the realization, and from there on out she is, indeed, looking at the world through changed eyes. The problem is, we don't know what happened that *finally* forced her to wake up and smell the damn coffee, and that's what we've been waiting for.

Here's how an ending like that feels: Imagine your best friend calls you, breathless, as she drives home from work. She says, "You know how my boss always yells at me, and I just sit there, taking it? Well, today he—" and suddenly all you can hear is every fifth syllable, and you know she's driving through a dead zone. Even so, you spend the next few minutes saying, "I can't hear you; can you hear me?" Until, finally, you hear her say, "... and that's when I realized that I *am* worthy and no one has the right to yell at me like that, so I quit." *Wait,* you're thinking, *what made you realize that?*

That is exactly what your reader is dying to know. Because this is where we find out what your protagonist is *really* made of. Which means that chances are it will tell us a little bit about ourselves, too. And that's why we're there, to gain insight into human nature, our nature—so we can better navigate the world.

Let Us Be *Inside* the Protagonist

The last layer to be aware of is the most crucial and the one that most often goes missing. In their zeal to show us what's happening externally, writers throw themselves into painting a vivid picture so we *see* the thing that causes their protagonist to realize his mistake. We're right there with him. But that isn't where we need to be. We need to be *inside* of him.

Otherwise, although we see the protagonist do something that lets us know that his worldview has profoundly shifted, the problem is, we have no idea why. Since external action is, by definition, external, the why behind what we do out here in real life is always tucked safely into our brain, where no one else can see it (thank heaven). In a story, however, the exact opposite is true: knowing why the protagonist does everything he does is precisely the point. Thus if we're locked out when your protagonist finally makes that big change, well, it's kind of like we made the trip for nothing.

For instance, let's say that at the end of a novel about Ankur, his father finally offers him the job he's lusted after since page one, *and Ankur turns it down.* Is it because he's realized he never really wanted to work for his dad in the first place? Is it because he's realized that he'll never be his own man unless he breaks away from the family business? Is it because he knows he's not really qualified, and is finally able to admit it? Is it because—ah, we could go on like this forever, but who has the time? The point is, we don't want to know *that* he's had a realization, we want to know *why.* The meaning, the emotional satisfaction, the inside intel, is in how he reaches the conclusion he draws as he makes sense of what's happening to him in the moment. That's what gives the reader a firsthand view of why he made the change, what it cost him internally to let go of his old belief, and what it gained him, especially since on the surface it might not gain him anything at all. In fact, it might *cost* him something. As in, Ankur now has no job and no income, and his view of who he is has been irrevocably shattered. Which might produce a surprising consequence for the reader to savor: Ankur has never been happier.

In other words, once again in literature, as in life, it's what's on the inside that counts.

Facing the End, with Confidence

In order to gather the information you'll need to write this very first, exploratory iteration of your "aha!" moment, there are three questions to ask.

At the End, Will Your Protagonist Achieve Her External Goal?

This is a straightforward question, and it relates to plot. It has to do not with how your protagonist will *feel* about whether or not he succeeds, but with what actually happens in the story. It's often an easy question to answer. For instance, Ruby's external goal is stated flat out in Jennie's opening sentence: Write or die. And the answer is that yes, Ruby will rewrite the ending of the script. For some writers, however, it can be a little more tricky, because although the *question* will be very clear—Will your protagonist live or die? Will they get divorced or stay together? Will ET make it home, or end up as a pitchman for Reese's Pieces?—the *answer* may be elusive. My advice, as with your opening scene, is to play it out both ways here, and see which one resonates most with you. Always keep in mind that the goal is to shove your protagonist as far out of his comfort zone as possible, the better for him to ultimately realize that it wasn't nearly as comfortable, or as safe, as he'd thought.

What Will Change for Your Protagonist Internally?

As T. S. Eliot said, "The end of our exploring will be to arrive at where we started and know the place for the first time."[6] So it's no surprise that at the end of the novel your protagonist will return, either literally or figuratively, to the place where she started, but now she'll see things very differently. That is, sans her misbelief. The question is, what will she have realized? In *The Wizard of Oz*, Dorothy starts out wanting to leave home at all costs, but at the end, when the Tin Man asks, "What have you learned?" her reply is, "If I ever go looking for my heart's desire again, I won't look any further than my own backyard. Because if it isn't there, I never really lost it to begin with. Is that right?" Well, in real life, probably not, but since that was L. Frank Baum's point, who are we to argue? After all, he's absolutely right about one thing—for better or worse, there's no place like home.

In Jennie's story, Ruby's internal realization will be that she had a far deeper connection with Henry than she'd allowed herself to see, and that instead of protecting herself from future pain, she was keeping herself from feeling genuine joy. She will also realize that—for better or worse— she is more connected to the world than she'd thought.

What Will Happen Externally in This Scene That Forces Your Protagonist to Confront, and Hopefully Overcome, Her Misbelief?

The "aha!" moment is the instant when your protagonist sees things clearly for the very first time, and her internal struggle is at last resolved, leaving her transformed (or, if it's a tragedy, not). That does not, however, mean that all internal conflict then suddenly evaporates, or that this is where the journey we've taken down the novel's third rail inherently ends. Especially in cases when the "aha!" moment is what *allows* your protagonist—who now sees things very differently—to leap that last hurdle, and face her external problem head-on. That is exactly what will happen to Ruby. Her "aha!" moment is what will give her the final insight—and the courage— to revise the script, and reveal herself to the world.

To zero in on your protagonist's "aha!" moment," you need to examine both pieces of your story's equation: internal and external. What will happen plotwise that will finally allow her to see her misbelief for what it is? And when it happens, how will she make sense of it, internally? Remember, we're not just interested in the conclusion she comes to. What we really want to know is, why did she come to it?

Here's Jennie working it out for Ruby:

> What Ruby wanted was to stay home alone in her house where she felt
> safe in the face of crushing grief and try to finish her script in peace,
> but the dog she snatched to try to accomplish that has forced her to
> leave home and be in the world—to be vulnerable (because she has no

protective mechanisms left), and to be trusting (because she has no choice but to trust some people—like Clementine, Nora, Sharon, and some random strangers who know things about dogs—in order to get what she wants).

That's when Jennie hit a snag, because since she didn't know what would actually happen once Ruby hit the road with the dog, it was difficult to envision what, exactly, might occur externally to spark Ruby's "aha!" moment. It was especially tricky because she had no idea why everyone was looking for the dog in her story, and how—specifically—that would cause problems for Ruby. That's when she stumbled onto what turned out to be the key that unlocked many layers of her story. Let's let her tell it:

> I knew the missing dog would trigger a massive dog hunt and decided that the dog *wouldn't* be owned by a famous person, but I couldn't figure out how to make those two things a logical reality. I was a little stuck, to be honest—and then something fortuitous happened: I was on a plane to Dallas, and I was reading the in-flight magazine, and there was an article about a woman who owned a famous Internet dog. The piece was about how she'd had to give up her job to become a famous-dog wrangler, and how insane it was and how much money she was making. I felt like the universe had delivered me a solution to my story on a silver platter! This was *it*!
>
> And so, at the very start of her story, driven insane with grief, Ruby would steal a dog who happened to be famous—and she would do so at a moment that was somehow important in the dog's life, which would explain the media frenzy and also explain why she would have to go on the lam with the dog. If she decided she needed to protect the dog from the harsh glare of the media (which she couldn't protect *herself* from), that would give her something to do—some action to take—rather than sit around wringing her hands and not writing her script.
>
> I still didn't know exactly who the dog owner would be, but I knew that *fame* was the thing that would connect Ruby and him. A shared understanding of exposure and identity would bind them together— and teach Ruby what she needed to learn. I had the feeling that this

knowledge is what would allow me to write a solid "aha!" moment scene for Ruby, and to finally "see" the entire novel.

I called the dog owner Tony and named the dog Rufus—and as I envisioned the scene, those characters both began to come to life in my mind. I began to see Tony as a guy who was having problems with love, too, which played off Ruby's misbelief very neatly. *This* was the moment when all the hard work really began to pay off for me—the moment when I hit on the famous dog thing, and so many other things fell in to place.

Note that Jennie spoke of that moment as fortuitous—because it really was. It was a stroke of good luck that she uncovered that nugget of information just as she was struggling with her "aha!" moment scene. Not to go all new-agey on you, but it's often the case that when writers are deep in a project, bits and pieces of helpful information begin to crop up, seemingly on their own. My take: It's because the more aware you are of the story problem at hand, the more likely it is that your trusty cognitive unconscious is on the lookout for anything that relates to it.

In Jennie's case, it was what allowed her to zero in on not only what would happen in Ruby's "aha!" moment, but why it would happen. Here she is, beginning to flesh it out:

> For the better part of the novel, Ruby's stuck with this high-maintenance dog whom she has to keep hidden. Things will get very complicated for Ruby because of the massive search for Rufus and the role of social media in it—which will mirror the role of social media in her writing predicament. Ruby's "aha!" moment will be triggered when she's forced to stop hiding and connect deeply with someone—the dog owner, Tony, who in addition to having some issue with fame, can maybe be looking at love in the wrong way, too. Ruby will see that she has made a tragic mistake in how she held herself back from Henry, and that realization will finally give her the ability to put her most vulnerable self onto the page and give the show a great ending, just before the novel's ticking clock runs out.
>
> As for where this big scene is going to happen? I have always pictured Ruby running away to a hotel. A fancy hotel—okay, yes, Lisa,

you and your why, why, why questions are in my head. Why a hotel? Well mostly because I think it would be so fun—kind of like that Tom Hanks movie where the guy was stuck in an airport—but *really* why? I know you're going to ask for that. Okay, it's because Ruby decides she needs to protect the dog—and herself—from the glare of the public eye. She needs to hide out. And she has no friends and no one she trusts and nowhere else to go. I can make sure this hotel has some kind of meaning for Ruby. Maybe it's where she and Henry used to go to get away . . .

Note that the same thing happened when Jennie worked on this scene sketch as when she worked on the sketch for her opening scene: she uncovered all kinds of questions she had no answer for. The scene's setting was just one—an easy one. She also has to figure out what Rufus and Tony's story is, so she can make sure it sparks her novel's third rail. These are questions she'll be answering soon, but for now Jennie has enough information to complete her Scene Card, and then take a stab at the very first version of the scene itself.

WHAT TO DO

Don't worry if you're still not 100 percent sure exactly what your protagonist's "aha!" moment is. Take a deep breath. Now, as best you can, answer the three questions we just went through. Remember, you're exploring here; there is no right answer. Take as much time as you need, and don't be afraid of writing too much, or even of dithering a bit along the way. Your goal here is to concretize the elements, both internal and external, that will then help you envision this scene. Like Jennie, you will no doubt uncover questions that hadn't yet occurred to you. If so, jot them down for future exploration. Once you're done, it's time to create a Scene Card for the "aha!" moment, and then take a stab at the very first version of the scene itself.

But before you do, let's push Jennie into the ring to see how it's done. Although there's still a lot of info missing, which we'll soon watch her uncover, she has enough to begin. First, let's take a look at her Scene Card.

SCENE #: RUBY'S "AHA!" MOMENT

ALPHA POINT: Ruby contacts Tony in order to give Rufus back

TONY SUBPLOT: Tony tells Ruby the truth about Rufus and himself; they bond

		CAUSE	EFFECT
		What happens	**The consequence**
THE PLOT		• Ruby contacts Tony, hoping he won't have her arrested; she just wants to return Rufus. • Tony doesn't want Rufus back. • Ruby doesn't want him either; they're at a standoff. • Tony opens up, and it turns out his story mirrors Ruby's—neither like the dog, both are struggling to make a human connection.	• Ruby comes up with a plan to help herself, Tony, *and* Rufus. • Tony will give Rufus back to the pound so they can use his fame to help other dogs find a home (which will free Tony to pursue the woman who left him because of how he treated the dog). • Learning why people love dogs so much helps Ruby understand why loving a person is so frightening.
		Why it matters	**The realization**
THE THIRD RAIL		• Ruby can no longer hide behind having to hide Rufus; she knows she needs to leave the hotel and face her life head-on, no matter how painful.	• Ruby's "aha!" moment is this: she finally realizes why Henry stayed with her all these years, and how profoundly connected she was to him—it's what kept her safe, and able to tap into the most real part of herself. She'd had the true love she'd always scoffed at, but never allowed herself to see it. • Ruby experiences crushing regret. • Ruby uses these feelings to fuel the script revision and apologize to Henry through what she writes.
			And so? Ruby rewrites the ending

Jennie then wrote the first draft of the scene in which Ruby earns her "aha!" moment.

I finally just did it. I went onto Twitter and searched for Rufus. I was stunned; he had a million more followers than when Nora had pulled up his profile, and it had only been two days.

I had 140 characters to play with, but I needed only 15.

"I have your dog," I wrote.

One moment went by, then two, and then the reply came: "Direct message me. Prove it."

I put the front page of the paper on the floor next to Rufus, so I could get in the date, then snapped a photo, making sure to show the customized collar—then I froze. I didn't know how to get a photo from the phone onto Twitter. I went back to the computer.

"Just took the photo with today's paper. I don't know how to upload," I wrote.

"Are you f-ing with me?"

"No. I don't know how."

He walked me through the steps, and when he had the photo, he called—the voice I had heard on TV and in video clips.

"Okay," he said, "Let's make a deal."

I slid down on the floor next to Rufus. "A deal? You mean like a ransom?"

Tony made a sound like the air shooting out of a balloon—a sound of disdain. "I'm not offering money," he said. "I'm offering the dog. I want you to take the dog."

I sat down on the edge of the bed, trying to take in this new information. I sniffed. "I don't understand," I said. "I don't want your dog. He's a diva and a pain in the ass. I actually never wanted your dog."

"The problem is that I don't want my dog, either," he said.

"*What?*"

"Everyone loves the fucking dog," Tony said. "They expect me to lavish love on him 24/7. It's gotten . . ." he paused, then lowered his voice. "Out of control."

"He likes fresh organic meat, doesn't he?"

"Yeah, and baths every night. I had a girl I really liked who walked out because the dog was whining and I told her to ignore it. She said it proved I was an asshole."

I wanted to laugh, but didn't. "Ignoring the whining is not a crime."

"I really liked this girl," Tony said.

"What is it about dogs?" I asked. "People are ridiculous about their dogs. The love they have for their dogs—I mean, I had no idea."

"I get it," he said, "I totally get it. Dogs can't hurt you. Safest love you can have."

I looked at Rufus snoring on the floor beside me, and thought about Henry lying broken in the hospital and of how much I must have hurt him over the years. I never let him get as close as he wanted to be. I always held back a sliver of myself—a reality that seemed, suddenly, to be the cruelest thing I could have done. Why had he stayed? Why had he let me do that to him?

"Yes they can," I whispered. "They can hurt you. They always die in the end." I pressed my lips together and let the tears roll down my face.

"Well that's fucking morbid," Tony said. "But the problem remains that I don't want the dog back. You're my ticket out of this nightmare."

My heart began to race. I thought of the cops at the dog park, and the girls with their cell phones. "You're going to pin this on me?"

"It's either me or you, sister."

I had visions of being thrown in jail, someone else writing the end of our show, never getting to say goodbye to Henry. "Wait, wait; I have an idea," I said. I didn't—I was just buying time.

"Shoot it to me."

"You need a way to end the Rufus show."

"Okay."

"I know something about shows ending."

"What is that supposed to mean?"

"Do you know the show *Complications Ensue*?"

"Yeah, I know it. Everyone at the gym watches it. I heard that one of the writers is on his deathbed and the other is choking on the last episode."

I closed my eyes, as if I could block the truth from getting inside. "It's me. I'm the writer. I'm the one who's choking," I said.

"No shit?" he said.

"No shit."

"For real?"

"For real. So like I said, I know something about shows ending. And what you need is an ending that gives everyone what they want."

"I don't follow."

"You said that people love dogs because they can't get hurt by them."

"Right."

"So you make an announcement that you've been hurt by Rufus. That you lost this girl you might have loved because of Rufus. And you explain that you can't be the owner of a famous dog anymore because you need to work on winning her back. Who isn't going to love that story?"

"She might not—the girl. Sue."

"Are you kidding me?" I said. "She'll love it."

There was a pause. "We can't exactly take Rufus back to the pound. People would skewer me."

I noted the use of *we*. "No, that's exactly what you *can* do. You give him back to the place where you rescued him from so that they can capitalize on his newfound celebrity and get massive free media exposure. They can have meet-and-greets. They can do photo shoots. They can get the endorsement dollars. It's the ultimate charitable gift—giving the pound back one of their dogs who became famous. It's a great story."

"That could actually work," he said.

"I'll call the *Times*," I said, "and the radio and TV people. You call the pound."

"You can't be there," he said, and there was something in his voice that was almost sad.

"I never met your dog," I said. "You can come down here and we'll do a handoff in the parking lot. Then you'll go to the press conference and tell people the thief turned himself in."

"No one will buy that."

"Did you ever read O. Henry?" I asked.

"Oh who?"

"The writer? O. Henry. Short story writer, early twentieth century, American. *The Gift of the Magi*?"

"Never heard of him."

"He wrote a story called *The Ransom of Red Chief.* These guys kidnap the son of a wealthy banker, and the kid is such a nightmare that they end up paying the banker to take the kid back."

"Are you calling my dog a nightmare?"

"Yes. Whoever took Rufus was going to demand a ransom but couldn't stand to be with the mutt. So they're giving him back. It'll prove your point about how hard it is to live with him."

"I like it," Tony said, but then he got quiet. "What if Sue turns me down?"

"Then you can start dating without a dog getting in the way."

"Girls only liked me because I had a dog," he said.

I shook my head, even though he couldn't see me. "Girls liked you because they thought you loved the dog."

He laughed. "I know," he said, "which made me feel like shit, to be honest. Like I was lying to everyone all the time. It's part of the reason I let myself get on this crazy famous dog treadmill. I thought if I accepted all the Rufus demands, I would feel better about myself. But you have no idea what those demands are really like. Fame is crap."

"Tony, I know all about fame," I said. "People say things about me all day long, and none of it is true and there's not a damn thing I can do about it."

"So you're not choking, then? The other guy isn't dying?"

I laughed for the first time in a long time. "Actually, he is dying. His name is Henry and he is dying. And I can't bear the thought of life without him. I can't. And I am choking so hard. The truth is that I can't figure out how to end my show in a way that will honor Henry and the story and not make me gag."

"You did a pretty damn good job coming up with an end to mine," he said.

"Thank you."

"And I have a confession to make, too, since we're getting all confessional. The truth is that I never loved the dog. I never did. I think Sue could smell the bullshit a mile away. She made me not want to be the guy who is full of shit."

"You know what Shakespeare would say about Sue? About what she did?"

"I can't say that I do," he said.

"*Romeo and Juliet*. Act 1, scene 3. *'Don't waste your love on somebody who doesn't value it.'*" Suddenly all I could think was *Henry, Henry, Henry*. He filled my head and the hotel room and the whole world around me. "I guess the opposite is true, too," I said quietly. "Don't ignore the love of someone who does."

Suddenly I had my answer to everything. Henry loved me because I loved the work we did together. He loved me in spite of my inability to love him back. He loved me in a way that was as true as you can get—and I had stood by and let it run through my fingers like water because I thought it would hurt too much in the end to give myself over to it. But here we were at the end, and I couldn't hurt any more than I did. I hadn't saved myself the pain. I had only cost myself the joy.

So that was how I would end the show. I would defy the conventions of tragedy, and my own original instincts, and let Juliet live. I would let her have the happy ending that Shakespeare decreed she would never get. Henry might never know it—at least not on this earthly stage of fools—but I would give him what he'd been asking for all along.

Tony and I made our plans for the dog handoff. I hung up and started to write.

That's a very strong scene, so it helps to remember that a big part of what made it so powerful was serendipity. Had Jennie not seen that magazine article, she might still be figuring it out. Point being: It may happen that you'll search for the perfect resonant moment and come up short—and that's okay. It's more than okay. It's the way writing works. Your "aha!" moment scene may feel a little thin compared to Jennie's, but as long as we're in your protagonist's head as your plot hits him with its best shot and he earns his "aha!" moment, you're set. You'll be circling back to this scene—and all scenes—again and again as you write forward, tweaking it, adding to it, layering it as you go. In chapters 12 and 14 we'll watch how Jennie begins to answer some of the questions her scene raised (about Tony, Rufus, why he's famous, and how it all connects to the novel's third rail), and you'll learn how to deepen your scene in the process. But first let's take a look at why this scene is so robust.

Why This Scene Works:

- Did you notice how, although the scene revolves around the external plot (Ruby and Rufus), what gives it meaning throughout is what Ruby's really working out here: why she's so afraid of really caring, and why Henry didn't abandon her? It's the internal struggle that drives it.
- Did you notice that Tony and Ruby have a similar take on the cost of human connection, and have in essence been struggling with the same thing?
- Did you notice, however, that each had approached the problem differently—Ruby spent her life running from it, while Tony was doing the opposite: trying to connect, but not allowing himself to be vulnerable enough to succeed? Each character therefore has the opportunity to learn from the other.
- Did you notice that in her conversation with Tony about Rufus, what Ruby is really doing is coming to grips with her misbelief? *It's not about the dog; it's about the cost of connection.*
- Did you notice that just about everything Tony said hit on Ruby's third rail, making her reflect on how she views other people (they're such fools about their dogs) and more specifically, her relationship with Henry (she realizes not only did he know her to her core, he accepted her as she was)?

- Did you notice that once Ruby has her "aha!" moment she realizes how to solve the overarching external story problem and rewrite the script?
- Did you notice that in solving Tony's problem—both the internal (the real reason girls like him) and the external (how to get rid of Rufus)—Ruby was also solving her own problem?
- Did you notice that when Ruby said, "I would let Juliet live," she is allowing her fictional characters to experience the full depth of the joy that she's been denying herself thus far?

WHAT TO DO

First, take a minute and review everything you already know about your protagonist. Reread every scene you've already written. Make a note of the memories your protagonist might call on during this scene as she struggles to makes sense of what's happening and what she should do about it. Imagine her boxed into a corner, with no choice but to take action. Or, if she *can* choose not to take a stand, then make sure that by opting out she would lose something she dearly wants—often the most meaningful, closely guarded thing a person has: their sense of self-worth. In other words, your protagonist has arrived in the land of now or never.

Now you're ready to develop your Scene Card and then write your protagonist's "aha!" moment scene. If you're wondering how the hell you're going to do this, you're in good company. This is scary, because there is no right or wrong. The goal here is to take everything you know will happen, merge it with what you suspect might happen, and then wing it! Maybe even write several versions. Sometimes the way writers realize that they want their protagonist to perish in the end is by first writing an ending in which the protagonist survives. Or, more commonly, vice versa. But here's the irony: it's only when you're in possession of the vast amount of info you already have that you *can* effectively wing it. Luck, as they say, favors the prepared.

11

BUILDING YOUR BLUEPRINT: HOW TO KEEP TRACK OF ALL THE MOVING PARTS

I admire anybody who has the guts
to write anything at all.

—E. B. WHITE

Now your novel is bookended. You have your opening moment—the last few seconds of your protagonist's "before," as life gets ready to toss him out of his comfy boat and into the waters of unavoidable conflict. And you have the "aha!" moment that will change your protagonist forever, allowing him to climb up onto the shore of "after."

At this point, you might be tempted to grab a stack of 3 by 5 cards and, thinking back to the external story structure models you may have used in the past, ask, "Okay, how many cards? How many scenes? When does the first act end? When should the climax come? And hey, do I have to make a card for every scene?"

Here's the skinny: neither your blueprint nor your novel is made up of a prescribed number of scenes, beats, turns, or plot points. As we know, story structure is the by-product of a story well told, not something that you can—or should—impose from the outside in. Your story will change, grow, shrink, and continually shape-shift as you write forward, finding its own organic architecture. Your goal is simple—build your story by creating a plot that will constantly force your increasingly reluctant protagonist to change. Put simply, the stakes will be ratcheting forever upward, even in those moments when it seems like the storm has already passed.

Throughout this process—indeed, throughout every draft of your novel from the first to the last—you'll continually be pinging from present to past and back again, refining what you've already discovered about your protagonist, and digging into other story-specific events in his or her life. You will find that the deeper you dig, and the more information you uncover, the more real all your characters will become, and it won't be long before you feel as if you're uncovering facts about them, rather than making things up.

What follows is the method you'll use to build your cause-and-effect blueprint—a method that is equally effective in helping you understand your story before you write page one as it is when you're writing forward, ensuring that each and every scene is there to serve your story. To be very clear, this isn't a process where you *first* blueprint your entire novel and *then* write it. Instead, you'll be writing, and *at the same* time developing the blueprint for what comes next—whether it's the very next scene, or scenes that will come into play much farther up the road.

It's in the Cards!

Now for the big question: *Do* you have to make a card for every scene? Yes, you do. Because even when it's a scene in which you're sure you know exactly what will happen, simply concretizing the specifics on a Scene Card first helps you keep your eyes on the prize. It'll save you from lunging after some beguiling (read: irrelevant) notion that will derail your story. Even small diversions are like the break in the rail that causes the locomotive to jump track and plunge into "who cares?" territory. What's more, distilling a scene down to the specifics forces you to answer these questions: Does this scene actually have a place in my novel's cause-and-effect trajectory? Does it follow from the scene that came before it, and does it lead to the one that comes next? It's a win-win.

You do *not*, however, have to finish all your cards before you begin writing your novel. My advice is to begin writing once you've fully fleshed out Scene Cards for the first five scenes, *in order*, along with the last scene—but as tempted as you might be to get cracking, please don't start working on these cards right now. There's still a lot to learn before you can use them to maximize your novel's potential. To that end, we'll spend the next few chapters mastering how to unearth the specific information that each card must contain, and in chapter 15 we'll turn to developing the cards themselves. No card can be fully developed until you've pinpointed the specific layers that it—and the scene it represents—will be built upon.

Remember, it's not about simply mapping out the external plot; it's about keeping the internal story and the external plot in balance, so each continually spurs the other. As we've already seen, when Jennie developed Scene Cards for her opening scene and her "aha!" moment, they immediately revealed places where more strategic digging into the past was needed before she could write the scenes in full. In this way the Story Genius blueprinting process will help you uncover the deepest truths you're writing about, and then help you get them onto the page.

Developing the first five Scene Cards may not sound like much, especially since you've already written your opening scene. But figuring out the four scenes that follow Scene #1 requires a lot of work, because this is where everything in the entire novel is set up, and just about all the balls put into motion. By the time you're ready to actually complete Scene Cards #2 through #5 (which we'll do together in chapter 15), you'll actually have dozens of other Scene Cards in various stages of development as well. In other words, you'll be well on your way to a complete novel blueprint.

It bears repeating that this is an organic and intuitive process, so at some times you'll be writing scenes, at others you'll be creating and developing Scene Cards, and sometimes you'll be doing both in tandem. Just about every scene you write will require a bit of digging into the past, and once completed, each scene will sow potent seeds into the future— seeds that will sprout into upcoming events, unavoidable conflict, and unexpected consequences. When that happens, you'll pause in the scene you're writing or the card you're developing just long enough to capture any future ideas on Scene Cards. This way, you're never writing into the dark, but always with a clear idea of where you're going next, the better to write a scene that actually gets you there.

And, as important, chances are those same seeds will also need to be planted in earlier scenes, either scenes that you've already written, or scenes for which you're developing a Scene Card. When that happens, you'll go back and strategically add in this new layer, making sure it plays forward.

There's one important caveat: Regardless of how much you jump around as you develop your Scene Cards, *you must write the novel itself in chronological order.* Writing scenes out of order is like building the sixth story of a building before you've built the second floor. Each scene not only leads to the next, but supports it, deepens it, and gives it meaning. The Scene Cards are how you develop that meaning, so that by the time you're ready to write a scene, it's already layered, and already part of your novel's cause-and-effect trajectory.

That's why it's so important to fight the urge to leap ahead and write scenes at random. Sometimes, I know, it will be impossible to resist—you'll wake up in the middle of the night burning with the desire to write a scene that won't happen for a hundred pages. That's fine; go for it. I wouldn't want you to explode or have your hair spontaneously combust (it could happen). Plus, sometimes that's exactly how gems surface, especially when you're trying to figure out how something that just occurred in your novel will play forward. That's why you wrote your protagonist's "aha!" moment scene—it made you focus hard on where your story is going, and it gave you something concrete to shoot for.

But making a habit of writing scenes out of order tends to result in scenes that may indeed be beautifully written, but that can't pull their narrative weight, not to mention threaten to lay waste to your novel's internal logic. This is especially true if you're writing speculative or historical fiction, in which you have to keep track of not only the external rules of the world, but also your protagonist's all-too-human logic (even if your protagonist is the last unicorn, a fuzzy footed hobbit, or *I, Robot*).

The Scene Cards will help you layer your scenes so each one has maximum power, urgency, and believability. They enable you to envision the multidimensional aspect of your novel in one fell swoop. Before long this back-and-forth layering process will become muscle memory—the natural way you approach every scene. When we talk about the 1 percent of writers who have a native genius for storytelling, *this* is what we're referring to; they have a built-in ability to do that multilayered envisioning without really trying. The good news is that the rest of us can achieve the same thing, one card, one scene, one moment at a time. And eventually, we'll arrive at the same place as those natural-born geniuses.

Now that you know how to build your story, the question is, how do you keep track of everything you're creating?

A Folder for Everything and Everything in Its Folder

Organization is important, especially when you're talking about something with as many moving parts as a novel. The good news is that, as we've been discussing, a novel revolves around one problem that complicates, so each and every part is grounded in the story logic you're creating from the very first page.

The trick, of course, is keeping track of all those scenes at different stages of development, along with the backstory that feeds them, and the characters who bring it all to life.

To do that you will create a set of folders—either by clicking that nifty magic "new folder" button on your laptop, or by actually heading over to the office supply store (on your laptop, of course) and having a nice big box of manila folders delivered. You might also consider using a software program developed specifically for writers, such as Scrivener, WriteWay, or Power Writer. Many of these programs would work well with the Story Genius system.

No matter which method you choose, you will need a folder for the following six categories, and you'll need a table of contents for each folder. The table of contents will grow and then shrink as scenes either advance to the next folder or are permanently booted once and for all. The folders are as follows:

- **Key Characters.** Every key character will have his or her or its (should it be a robot, cyborg, or Brave Little Toaster) own folder. This is where you'll place their story-specific bio (which we'll discuss in chapter 14), along with any backstory scene in which the character appears. For scenes in which two or more characters appear, put a copy into each character's folder. As these folders grow, they'll become a font of inside info from which you'll constantly be drawing.

- **The Rules of the World.** If you're writing a novel that unfolds in a world different from the one we wake up to every morning—think sci-fi, horror, magical realism, futuristic, historical fiction, and all things speculative—you'll need to keep a keen eye on what is possible in the world you're creating, what is patently impossible, and most importantly, why. Simply put: This is where you will keep track of the logical framework in which the world is grounded. After all, it's the world your protagonist not only will have to deal with but has most likely grown up in.

- **Idea List.** This is where you'll put ideas that are still too fuzzy or too conceptual for you to envision as an actual scene. Things like: *Raul feels guilty about lying to his parents* or *Aisha struggles at work*. You have a vague notion of the kind of thing that will happen, but without any specifics, it remains completely conceptual. In other words, it doesn't have an Alpha Point. Keep this list in chronological order based on when you think it would happen in your novel, so that new ideas don't simply get tacked onto the end of the list. Yes, even at this stage, you're thinking cause and effect. In the beginning a lot of what you know about your novel might very well be on this list. As soon as you've concretized an idea, whisking it from the realm of the general into the land of the specific, it leaps into one of two scene folders:

- **Random Scene Cards.** This is where you'll put any scene that you can actually envision—which means it must at least have an Alpha Point—but that doesn't seem to have a connection to your novel's third rail, and/or to your novel's external cause-and-effect trajectory. Any romping around in the story goes into this folder, and many cards will never make it out. For a card to advance from this crowded catchall file—aptly known as the Random Folder—into the far more exclusive Development Folder, it will have to prove its worth by becoming relevant to the story you're telling. Only the best and brightest will earn entry. Here's how they make the leap:

- **Scene Cards in Development.** This folder is for cards that you've determined *are* story specific and so are eligible for a place in your novel's cause-and-effect trajectory. Not all of the cards in this folder will make it into your novel; some will be discarded along the way as your story expands and takes shape. But because all of them are story relevant from the get-go, none will have the power to wreak havoc from the inside, derailing your novel's finely honed inner logic. This is where your actual blueprint lives. These cards will be arranged in a rough chronological order, scene by scene. Once a card enters this folder, it gets a scene number, based on where it falls. Naturally, these numbers will change as you add or subtract scenes. But, and this is the point: numbering your scenes will always keep you focused on where each falls in your novel's cause-and-effect trajectory. When you're not sure exactly where a scene will go, put it at the front of the timeline, so you'll be aware of its pending presence every time you look through the cards. It will take time to fully develop each card, and there's no need to rush. Once a card is complete, it will be your guide as you write the actual scene.

- **Scenes.** Ta-da! This is where your manuscript lives. Right now, it's where your opening scene and your "aha!" moment scene are.

WHAT TO DO

Take some time and set up your folders. I'm betting you have more to put into them than you might imagine, and chances are as you fill them you'll begin to see a burgeoning synergy among the scenes, characters, and backstory you've been building. Plus, there's real satisfaction in organizing what you've already created and seeing it all together in one place. Take a moment to savor it.

How do you find the info that will populate those folders? It's kind of like what Deep Throat told those *Washington Post* reporters who were tracking down the Watergate burglars in *All the President's Men*: follow the money. Except instead of money, you're tracking the "why?" of your story. You're going to follow every "why?" that pops up—and as you'll see, every question you answer tends to give rise to a handful of other questions, so you're going to encounter a lot of "whys." It's a messy process, and it might feel chaotic at first. You'll be sniffing out information on myriad layers, going in several directions at once. What will keep it from actually *being* chaotic are the Scene Cards. They will allow you to grab, concretize, and organize that information, not *in general,* but specifically, and—this is the real bonus—in the order in which it will appear in your novel. So don't worry if it feels a bit overwhelming. Leap into it with reckless abandon, and revel in the messiness of it!

12

GOING BACK TO MOVE FORWARD: HOW TO HARVEST THE PAST TO SET UP THE PLOT

Actions are the seed of fate; deeds grow into destiny.
—HARRY S. TRUMAN

Armed with a backstory that delivers your protagonist to the opening scene of your novel, it's time to begin building an escalating plot that will force the protagonist to earn his "aha!" moment. This will help you begin developing not only your first five Scene Cards, but many more. Thanks to the work you've done and the scenes you've written, chances are you already have several specific plot twists and turns in mind. Not because they're dramatic, surprising, or insightful in and of themselves, but because they organically stem from your protagonist's internal struggle, which is what *makes* them dramatic, surprising, and insightful.

But at this stage, even though you probably have a general idea of what will happen in your novel, those concrete plot points are few and far between. That's because right now you know more about your protagonist's past than you do about his present or future, so just like in real life, while your vision of his future seems totally clear and doable *in general,* the moment you take a good hard look at it, it tends to vaporize into a mist of "Wait, what was that again?"

In this chapter we'll bravely wade into the fog and begin identifying specific scenes and plot points by sifting through what you already know about your protagonist's past, so you can begin creating a surface plot-level cause-and-effect trajectory driven by his internal struggle. It's a trajectory that will never stray from the single, complicating problem that your novel revolves around, because what propels it forward aren't random, external events, but the power of your protagonist's reaction to what happens. By the end of this chapter, you'll have started several Scene Cards, each of which will capture specific information you already have and reveal specific information you'll then begin digging for.

Don't Let Your Plot Run Away with Your Story

Believe it or not, even with all the incredible work you've done, it's *still* astonishingly easy to let the plot bully you into hitching your novel's momentum to a random external trajectory, as if the protagonist's past was merely the bus that delivered her to the current problem and then drove off in the opposite direction.

This is a trap both novice and experienced writers fall into, and is precisely how they end up with one of those one-size-fits-all plots. For instance, once Ruby goes on the lam with Rufus, it would be completely understandable if Jennie stumbled into this type of very seductive trap (not that we'd let her, mind you). Here she is riffing on all the crazy dog things that could happen once Ruby and Rufus are on the run:

> Ruby could run into a pack of crazy dog lovers who are spearheading the search for the famous dog, and they could be organizing search teams and holding press conferences about the rights of dogs. Maybe they try to rope her into leading a search team, and because she's so reluctant they get suspicious of her, and now she has people searching for *her*.

Sounds great, right? There's a very clear external cause-and-effect trajectory at work here; it's almost as if the plot were writing itself. And that's what makes this so seductive. In and of itself, it makes total, complete sense. Until we ask, *What the hell will any of this have to do with Ruby's actual problem, the one she entered the novel with?* That's when the story-smothering flaw of this approach emerges. This plot jumps track the minute Ruby gets pulled into the search for the dog. Why? Because this novel isn't about a dog caper, or the world of dog lovers; it's about a woman who has three days to write her magnum opus, while coming to grips with the fact that her fear of loss has robbed her of the one thing she most wanted: genuine, deep human connection. What does that have to do with the high jinks that will ensue if Jennie unleashes the dog caper scenario and lets it romp playfully all over the story she's trying to tell? Nothing.

Does that mean one of us had better tell Jennie that it turns out she really does need to eighty-six the whole dog thing? Of course not! Instead, the question we need to focus on is, how can Jennie build a plot around the dog caper that will help force Ruby to confront the internal problem that spurred her to kidnap Rufus in the first place? A plot that will continually nip at her heels, dogging her steps, relentlessly hounding her into resolving her internal issue one way or the other? (I know. Sorry. It was fun, though.)

The answer is surprisingly simple: by always keeping her eyes on the prize—Ruby's internal struggle, and whether or not she'll be able to write the script on time—Jennie will be able to do just that.

It's the constant laser beam focus on your protagonist's story-specific inner struggle that will keep you from allowing surface storylines to hijack the story you're telling. This is crucial, because the minute we race after that bright shiny thing—the enticing lure of a storyline that gallops off in its own direction—it's over. In the same way that meaningful specifics beget meaningful specifics, so do meaningless specifics lead to more of the same, which is why once a novel is pulled off the rails, it's really hard to get it back on track. Plus, it can mean ditching months of work. Because all those "meaningless specifics" that the first misstep led to? They're part of a chain reaction that can keep going for hundreds of pages. Once you realize your mistake and trace it back to the original offender, they *all* have to go.

That's not to say that Jennie might not end up with some of these same plot points, but if so, she will develop them with a very different focus in mind. To do that, she will ask of each plot twist, *before* she lets it anywhere near her blueprint, how will this event drive Ruby's internal journey and so propel my story forward?

Using the Past to Divine the Future (aka the Plot)

To keep your novel on course right out of the starting gate, we're going to build your blueprint, based on the information you've already uncovered, and begin to capture it on Scene Cards. There are two steps: first, we're going to review everything you've written and harvest the plot points that have already surfaced—that is, the ones you're already aware of. Next, we're going to dive deeper into the three scenes you wrote in chapter 7 to pinpoint what else, with a little more burrowing, might be right there waiting for you to uncover it.

So take a minute to look at all the work you've done thus far—from your *What If* all the way up to your "aha!" moment. Feel free to pause to savor the fact that even though you haven't developed your plot yet, you've already written large sections of your novel in the form of flashbacks and the memories that your protagonist will use to make sense of everything around him. And you've done something else too. While digging deep for your story, you set a lot of events in motion that are going to roll through your novel, ideally wreaking havoc for your protagonist.

In fact, by now chances are you have an idea of what will happen in your novel. Not so much the nitty-gritty specifics, but the general shape of it has probably been kicking around in the back of your head throughout this process.

Here's what Jennie wrote:

> As Henry lies dying, Ruby is forced to revise the final script of their
> hit show alone—and she starts to come unglued. When Nora, her

veterinarian sister, realizes the extent of Ruby's despair, she threatens to physically remove Ruby from her house. Ruby's plan to get Nora to leave her alone is to steal a dog for one afternoon, to prove to Nora how grounded and caring and on the ball she is—but the plan goes terribly awry when the news of the dognapping goes viral and Nora realizes that not only has Ruby stolen the dog, but there's a massive search on for him. Now dog devotees are frantically hunting for the dog. Meanwhile, totally separate from the dognapping, fans are threatening to take over the show that is Ruby's last connection to the love of her life. Trapped in a hotel with a diva dog she can't let anyone see and a looming deadline, Ruby has to confront the wild and irrational love people feel for dogs—and for each other.

Although this is pretty general and a tad clunky, it's a promising start, because each "event" evokes a strong reaction from Ruby based on the subjective meaning she reads into it. For instance, Nora's threat *causes* Ruby to panic, which is *why* she decides to nab the dog. Your goal is twofold:

1. Make sure each event causes the next one to happen, in an escalating succession as things go from bad to worse.
2. Tie each event to the internal change it triggers in your protagonist, giving a glimpse of why, and how it then triggers the next thing that happens.

This will keep you from ending up with a long list of events that are nothing more than a bunch of things that happen.

WHAT TO DO

Take a minute to jot down a brief overview of your novel as you see it right now. This is similar to what we did back in chapter 3 when you wrote out your *What If,* only this time it's far more specific, far more developed— and a definite indicator of how far you've come. Think of it as back cover copy, offering up the plot problem your protagonist will face, how it will escalate, why it *is* a problem, and what it might cost her, emotionally, to solve it. It doesn't have to be well written, and you never have to show it to anyone else. It can be as long or as short as it needs to be.

Don't worry if your sketch is just as bare as Jennie's. As we're about to see, a sketch like this is the literary equivalent of one of those circus clown cars. You'll be surprised how many potent plot points can leap out of one short paragraph. The caveat is that this only works because you've already spent a good bit of time envisioning what has caused the events that will take place in your novel. In other words, your paragraph isn't "from scratch" but drawn from material you've already developed.

How to Look for Trouble (for Your Protagonist, That Is)

Now, just like in chapter 8 when you were searching for your novel's overarching plot problem, take a look at your sketch and see if you can pinpoint all the moments that challenged your protagonist and caused her to take action. Make a list of every potential scene, plot point, and storyline that springs from your paragraph, so you can continue to develop them.

Here is what instantly leaped out of Jennie's overview:

- Ruby will try to write her show—and be thwarted at every turn by external events, grief, sorrow, and Rufus, the diva dog.
- Ruby's sister, Nora, will recognize Rufus, and instantly realize what Ruby has done.
- Ruby will have to deal with Nora, who is beside herself that Ruby has taken off with a stolen dog, and worried sick about Ruby's mental health. I'm thinking that something might happen to Rufus that Ruby will need Nora the veterinarian to fix—something that's a result of Ruby's neglect or cluelessness about dogs.
- Ruby will have to take care of Rufus—feeding him, walking him, actually seeing to his needs—something she's never had to do for any member of any species until now.

- Because of the publicity from Henry's accident, and the fact that the spotlight is now firmly pointed at her and only her, Ruby's life will be dissected, commented on, and analyzed by fans, something she'll have to deal with internally, and ideally for the story, at times, in person.
- Ruby will have to deal with her jealousy of Clementine, the fangirl, and the script Clementine has presented as a viable alternative to Ruby's nonexistent work.
- No one (besides Nora) knows that Ruby has Rufus, but someone will begin to figure it out and close in on her—ohh, maybe it could be Clementine herself?

Everything on this list brims with *potential* action, meaning, and possibility. But, as you can see, just about everything on it is also general. The vast majority of these bullet points are too vague to yield a scene at the moment. They're still in the idea stage—we know, in general, the kind of thing that will happen, but we don't know any of the specifics. For instance, how does Ruby deal with her jealousy of Clementine? Close your eyes. Can you picture it? Nope. Thus it's still too conceptual to begin a Scene Card. So Jennie added each potential scene to her Idea List—except one: "Ruby's sister, Nora, will recognize Rufus, and instantly realize what Ruby has done." It's the only single, clear-cut event in the bunch, so Jennie started a Scene Card for it.

WHAT TO DO

Go back to your overview and make a list of any potential plot point that springs out at you. I'm betting that as with Jennie's, the vast majority of them are great, but general. Add them to your Idea List, and start a Scene Card for anything specific by filling in whatever pertinent info you've discovered—but, as always, don't worry if it's only the Alpha Point for now.

Discovering What Can Go Wrong, So You Can Make Sure It Will

Now that you've harvested the obvious plot points and challenges your protagonist will face, you're going to build your cache of potential scenes by combing through the turning point scenes you wrote in chapter 7 in search of other, less obvious plot points that might surface to haunt your protagonist, hopefully when it will cause the most conflict. Chances are you'll unearth several gems there, since those scenes focused on specific moments in your protagonist's past when his misbelief guided the hard choices he made. All of those choices had ongoing ramifications, many of which will come to fruition in your novel. Your goal now is to find them and either add them to your Idea List or get them onto a Scene Card. There are two main areas where potential plot points lurk.

What Secrets Does Your Protagonist Have, What Lies Has She Told—to Others, and Even More Importantly, to Herself?

Secrets and lies go hand in hand. After all, secrets are often the flip side of a lie, as in, I told a lie (of course I paid the electricity bill!), because I wanted to keep a secret (I lost my job and can't pay any bill). Not to mention that an unexposed lie is, by definition, a secret—that's kind of the whole point of lying. And when it comes to secrets and lies, out of sight is *not* out of mind. For instance, Jennie envisioned a scene where Ruby purposefully torpedoed her impending marriage to Henry on their actual wedding day. That's clearly not only something that might have troubled Ruby since then, it's also a big fat secret lodged in there like a time bomb, waiting to be set off.

But in order to light the fuse, Jennie would have to figure out what, exactly, Ruby *did* to sabotage her wedding. That's the beauty of searching for the secrets and lies that follow your protagonist into the novel—it forces you to concretize them by making them specific, which is the only way they can play forward.

So I asked Jennie what, specifically, Ruby did back then. Jennie knew that the easy answer would be that Ruby orchestrated some scenario wherein Henry caught her cheating on him. Jennie sat with this idea for about three seconds before turning her back on it. It didn't ring true to her for this story, or these characters. And were Ruby to do such a thing, it was even more impossible for Jennie to imagine why Henry would still love her afterward—an integral part of her story. She joked that thinking about it was giving her a headache—and that gave her the answer:

> She can have a migraine! Back when I was first figuring out how Ruby got out of marrying Henry, I had a general notion that stress did her in. But what if she actually makes herself violently ill, on purpose? There can be an unusual weather event on this day, which is often a trigger for getting a migraine. All Ruby needs is to invite in a second trigger, and she knows she will have actual, debilitating pain, and the wedding will have to be canceled. I know this truth from my own life—and I have actually already written a character in another novel who did something like this. What can I say? As a migraine sufferer, I am intrigued by the concept that I am a constant danger to myself. I have never used this power in my own life, but I sometimes imagine that I could . . . so I will give Ruby this power, and I will allow her to wield it in secret, and then conceal the fact that she has done it.

This just leaves one unanswered question, which I posed to Jennie: why don't they just reschedule the damn wedding? This time the answer came to her instantly.

> Because Henry has never seen Ruby so visibly ill and debilitated—I mean, she's throwing up, feeling faint, moaning in pain—he feels horrible for pushing Ruby so hard. He knows that she put her feelings for him ahead of her fear of marriage, and now it has made her sick. So he tells her they can go back to the way they were—and they do.

Excellent! And did you notice that this actually yields two Scene Cards? The aborted wedding scene (backstory, yes, but you *know* it will be in the novel), and a scene when Ruby's deceit surfaces. She won't be able to admit it to Henry, but perhaps to someone else, another character who was affected by it, and whose reaction might give Ruby fresh insight. So Jennie started Scene Cards for both and put them into her Development Folder.

WHAT TO DO

In culling through your protagonist's specific past, you will no doubt find several similar possibilities, some of which you'll use, and others that you'll stash in your back pocket, just in case. And by back pocket, I mean you'll start a Scene Card for it and slip it into the appropriate folder.

Add any general "maybes" to your Idea List of almost-theres.

What External Obstacles Have You Already Planted in the Past That Will Keep Your Protagonist from His Current Goal—or Conversely, Help Him Attain It?

For most of you, this category will yield a lot, including promises, agreements, or contracts your protagonist made; people your protagonist betrayed or wronged; and long-standing feuds between family members or friends. Given what Jennie knows of Ruby's past, *as yet*, she doesn't have a lot of specific external obstacles to draw on. The key phrase here, of course, is *as yet*. Because there is plenty of potential—telltale events in Ruby's shared past with Nora, Clementine, and Sharon will yield grist for the mill as Jennie continues to expand her plot and flesh out her Scene Cards. Like you, she'll continually be digging into the past, to discover the origin of obstacles she hasn't yet envisioned.

And that's where we were going to leave it. But I kept pushing, and then out of the goodness of her heart (and, yes, to humor me), Jennie agreed to go back and look over what she'd already written one more time, just to see what else might come up. That's when she may have hit the jackpot.

- What about Beth? I haven't given her much thought since the skating rink incident. There might be something there—even if Ruby hasn't spoken to Beth in decades, the Internet is the great connector. Beth might pop up and help Ruby in some way, maybe by giving her fresh insight about the past. Maybe she'll trigger Ruby's realization that she read the situation wrong back when they were kids.
- Ruby could run into someone at the hotel where she is hiding out— maybe someone really inconvenient for her, like someone from her Hollywood world. Another writer or an actor. Oh! It could be the producer who hired Henry to write the movie script and left Ruby out. He could be the face of rejection for Ruby, plus he's the guy who ruined everything. If he hadn't tempted Henry away, none of this would have happened.

Good ideas, both of them, but I spotted something else that leaped out of Jennie's overview that had real obstacle-producing potential: the massive search for Rufus. Since a lot of Jennie's plot will revolve around Ruby hiding Rufus, it will clearly be a source of obstacles she'll have to overcome. But in order to figure out what they might be, there is one question Jennie had to answer first: Why *exactly* will there be a massive search for Rufus? What is going on in the dog's life (and Tony's life) at this moment that's triggered the search, and who exactly is searching? Here's Jennie digging into that question:

In the article I read, the dog's owner was complaining about what a hassle it was to cater to her dog for all his appearances. She quit her job in order to get Tuna (that was the dog's name) to all the photo shoots and sit for all the interviews. At first that would probably be fun—entertaining, a lark. But after awhile? It would begin to wear on you. And you can't hire someone else to take the dog to these appearances—the talk show hosts and reporters need the owner to speak for the dog. So there's no way out. The dog becomes your career, your identity—and the whole world is watching your every move, hanging on your every action, so it's not like you can just quit and do something else. I mean, once a Kardashian, always a

Kardashian, right? So what if at one point Tony was into the famous dog thing (for some reason that I will come up with), but now he's at the end of his rope? And what if it's right at the moment when there is some big demand on Rufus—a particular appearance or launch of something? It could even be that Tony was hoping against hope the dog *would* get taken.

From there it was easy to begin spinning what could happen as a result, enabling Jennie to begin to envision plot points that might actually have a place in her novel:

What if Tony has a deal with a dog food company and Rufus is supposed to show up for a big photo shoot with some famous actor, so the dog food company has put up a reward for the person who finds the dog? Or maybe everyone on social media is insinuating that whoever took Rufus did it for ransom, and so the consequences of Ruby being caught are ramped up. There are so many possibilities! What's funny is that I was winging this—just really riffing and thinking of the most outrageous things I could imagine. I didn't think I'd come up with anything I would actually use. But the second I went back to that magazine article, it all began to come clear.

This is something that I've seen happen over and over—when a writer lets her imagination loose *within the context of her story* it not only uncovers gold but also tends to beget more questions. Point being: Very often when you find the solution to one plot hole, it brings with it a few more holes of its own. For instance, if Rufus *is* that famous, why doesn't Ruby recognize him herself?

Jennie almost panicked, until she reminded herself that Ruby shuns social media of all stripes, and has left all things digital to Henry. In other words, Jennie discovered she'd already solved the problem. Coincidence? Maybe. Or, far more likely, because she's so focused in on her story, plot points have started appearing on their own. Not by magic, but because, given who Ruby is, that *is* what she'd do. Often when reviewing what you know about your protagonist's past, you'll discover answers to questions you hadn't yet asked. How great is that?

WHAT TO DO

Chances are you uncovered several possible obstacles that will thwart your protagonist in her (oft inadvertent) quest to overcome her misbelief. Either add them to your Idea List, or, if they're clear enough, begin a Scene Card for each obstacle you ran across. Do not be discouraged if you have very little to put on any of your cards at this point. As we know, scenes, characters, subplots—indeed, novels themselves—are built one layer at a time, which is why it will take awhile to fully develop each Scene Card. In the next chapter we'll return to the past to dig for the specifics that will transform the general plot points you've unearthed thus far into vivid images, concretizing them and bringing them into sharp focus. To do that we will once again whip out that most potent of writers' tools: the deceptively simple, all-powerful question, why?

13

STORY LOGIC: MAKING SURE EACH "WHAT" HAS A "WHY"

Truth is stranger than fiction, but it is because Fiction is obliged to stick to possibilities; Truth isn't.

—MARK TWAIN

You now have a promising—albeit foggy—notion of some of the things that might happen in your novel. Since a lot of it is still very fuzzy, at the moment your Idea List is probably much longer than your list of fledgling Scene Cards. Don't be discouraged—at the moment, just about everything you have is relevant to the story you're telling, so you're already miles ahead of the pack. Now we're going to begin the process of turning those broad generalities into the compelling specifics.

How? By asking "Why?" Because you can't know what, specifically, will happen until you know why it will happen. The "Why"—the reason something might happen, can happen, does happen—is what creates your novel's internal logic, so that things add up, and your reader can eagerly anticipate what might happen next.

Asking "Why?" is what burns through the fog, allowing you to envision your story's cause-and-effect trajectory—clear, precise, and waiting for you to bring it to life.

And the answer always lies in the past.

Take a look at every scene you're considering, and you'll notice there are two basic situations that will send you into the past to do a bit more digging. The first is when you have a specific plot point or scene that does not yet have the logic to back it up and make it believable. The second are those large swaths of territory that you've summed up in general (and in general, they sure sound great!), but that lack any real specifics, so when you try to envision what would actually happen, they become terrifyingly blank.

In this chapter we'll take a crack at one of each. Our goal? To master wielding "Why" in order to uncover the specifics that will crystallize into genuine, flesh-and-blood scenes.

Testing the Logic of Existing Plot Points and Bringing Them to Life

Chances are the first fuzzy scene on your Idea List is something that happens very early on. For Jennie, it was a rather all-encompassing plot point, the moment that sets up the entire cause-and-effect trajectory that sends Ruby on the run with a stolen dog.

Here it is, straight from Jennie's Sixth Tick: Ruby's veterinarian sister, Nora, is so worried about Ruby's state of mind that she plans to force her to leave her house and spend a couple of months in Ojai. Ruby will take this threat so seriously that she decides to "borrow" a dog to convince Nora to leave her alone. Sounds fine, right? (You know, other than that it's proof that Nora has a pretty good point about Ruby's mental state.) It's clear, it's concise, it's concrete, you can even picture it.

But here's the problem, as Jennie noted when she came up with this plan: there is a great big assumption lurking in that otherwise straightforward sentence. Since Ruby is a successful TV writer, and has yet to do anything that would make her genuinely certifiable, what would give Nora the power to make her leave her home?

While Jennie saw this one coming, the truth is that writers often don't. Instead, once they get what seems like a good idea, they run with it as if it's already a given. And so they gallop forward on the back of a plot point that sounds perfectly plausible, only to discover pages later that it has a logic glitch that renders it—and everything they've written since—completely moot. That is not a good writing day.

While letting go of ideas that seem so promising at first blush can be painful, it's far easier at this stage than it would be later, when you've already become so accustomed to the idea you've been developing that booting it would be like firing your best friend from the job she's wanted all her life, even though she has no aptitude for it, and everyone else in the office is threatening to quit unless you do something about it.

That's why it's essential to test every plot point for believability right out of the starting gate. The beauty of this is that by digging for the "Why" you also uncover more of the "What" that you need to move the story forward. By now your mantra should be, *specifics beget specifics*.

So Jennie set out to discover why Nora has the power to boot Ruby out of her house. To do that, she ran her fuzzy plot point through three "Why" tests, each more difficult than the last.

Why Does My Plot Need It to Happen?

This is the easiest question to answer, and it will help you nip a lot of tempting, but ultimately random, plot points right there in the bud. Jennie's plot point clearly passes this test, since her entire plot hinges on Ruby's snatching the dog. So she moved on to the second layer.

Logistically, Why *Can* It Happen? In Other Words, Is It Actually Possible?

Yep, Jennie's fuzzy plot point gets snagged on this one—there is clearly a logistics problem. It's not that what she's proposing is inherently impossible, like "and then Nora sends Ruby back in time so she can sample the Paleo diet for real." (Of course, if Jennie were writing speculative fiction, then her job *would* be to make that believable.) Instead, Jennie's job is to make something that *could be* possible—Nora's threat to drag Ruby out of her house—into something that *is* possible given the specifics of the story she's telling.

She immediately began brainstorming dramatic possibilities:

- Nora owns a stake in the show's production company or is somehow part of management, and Ruby's house is actually owned by the company.
- Nora is a shrew, and Ruby is docile.
- Nora is a hypnotist; all she has to do is dangle a pocket watch in front of Ruby and she'll do whatever Nora wants.

Clearly, that last one was Jennie's frustration peeking through. The trouble with all of these solutions is that none of them dovetail with who these characters are, and none touch the novel's third rail. That's when Jennie started to wonder if perhaps she was overthinking this whole thing, and took refuge in Occam's razor: the simplest solution is usually the right one. What if Nora owns the house Ruby lives in? That way if she were worried enough about Ruby's mental health, she could force her to leave.

Mission accomplished! Now the plot point is logistically possible. The next step is to drill into why Nora owns Ruby's house, one layer at a time, to see if this scenario now meets the last "Why" test.

Why Would It Happen, Given Your Protagonist's Inner Struggle?

This is what gives you insight into the plot point's true meaning. As you may have guessed, once a plot point has cleared the first two "Whys," the answer to this question will determine whether it belongs in your novel.

To figure out whether storywise it made sense for Nora to own Ruby's house, I posed a series of questions to Jennie:

LISA: Why does Nora own the house?

JENNIE: Because their parents willed it to her.

LISA: Why did they will it to Nora and not Ruby?

JENNIE: Because Nora is much older than Ruby. Ruby was a child when they drew up their wills, and they never bothered to revise them. Problem solved, right?

LISA: That's a good start. It's believable. But it's still just logistics. How does it fit into Ruby's misbelief?

JENNIE: Ugh! Okay, fine. So if the will is in play, that means the parents are dead. Maybe they died when Ruby was still pretty young—just grad-uating from college, a seamstress trying to break into the film/costuming world—and having a house in L.A., free and clear, was a godsend. I can imagine that Nora would've been eager to reconnect with Ruby—they hadn't been close since Nora left for college when Ruby was twelve, just before Mr. Anderson died. Ruby accepted the offer of a place to live, but kept Nora at arm's length emotionally. This hurt Nora, because taking care of Ruby was the only experience of family love she'd had growing up. Without that, Nora would've had to face the fact that her parents were as disconnected and cold to her as they were to Ruby. So out of her own fear and longing, Nora basically gave the house to Ruby, but never in any legal way. Nora has never lorded the home ownership over Ruby, never made a threat about it—until now, when her concern for Ruby is overpowering.

Good! As you can see, the deeper you dig into what began as a sim-ple logistical question, the further you get from logistics. Jennie is now heading into the heart of something far more internal, and far more intriguing—the subjective "Why" that drove Nora's action (allowing Ruby to keep the house), and the reason this plot point will definitely spark the novel's third rail. Nora, it turns out, actually loved Ruby whole-heartedly, and Ruby loved her, too—that is, before her misbelief took hold and walled her off from it.

As Jennie filtered this fuzzy plot point through all three layers of "Why," it's gotten more specific, and far stronger for it. Her goal was to make sure her existing plot point made sense, both externally and internally—done! While there is nothing for Jennie to add to a Scene Card at the moment, the gain is even greater: she can not only keep her opening scene as is, she can continue forward with this storyline, knowing it's not only logistically possible, but psychologically viable.

But wait, there's more. In addition to giving this plot point the power it needs to drive the story forward, Jennie also uncovered a bit more "What." Sure, each nugget is as fuzzy as the original plot point, but each one gives her an idea of where to dig in Ruby and Nora's shared past. For instance, here are three juicy areas she can explore:

- Nora and Ruby had a close relationship before Nora went away to college, and given that their parents were so cold, that probably left a big hole in Ruby's life.
- Ruby *did* love Nora wholeheartedly before Nora left and Mr. Anderson died, which means that Nora might be the only person Ruby ever really gave her heart to. So actually Ruby felt abandoned when Nora left. Ah, yes, abandonment issues!
- Nora has the power of legal home ownership over Ruby—which Ruby could think was a moot point. She could think she had a sort of "common law" ownership of the house—you live in it long enough, it's yours. As far as she's concerned, the house is hers.

Since the sisters' relationship definitely plays a major role in the novel, Jennie immediately added everything she'd uncovered to *both* Ruby's and Nora's bio and slipped it into each character's folder.

WHAT TO DO

Now it's your turn to scan your Idea List and Scene Cards, hunting for plot points where the logic and/or the logistics are still fuzzy—and don't be surprised, or disheartened, if it's all of them. Explore them in chronological order. This is important, because as with Jennie's foray into the relationship between Ruby and her sister, the info you uncover as you probe that first plot point will, by definition, affect, clarify, and give you insight into your second plot point, and so on throughout. The goal is to make sure each fuzzy plot point can pass all three "Why" tests.

When this leads you to fresh info about your protagonist, or any other character, make a note of it and put it in that character's folder. New fuzzy plot points go on your Idea List; then, should any specific scene leap into being, begin a Scene Card for it and slip it into either your Random or Development Folder.

Filling in the Blanks: Creating New Plot Points

What happens when you hit one of those large empty swaths in your cause-and-effect trajectory that don't have any clear plot points at all? Think: the sections that you've summed up in general, but that lack any real specifics. Things like "Kenisha will suffer a series of setbacks at work," except you don't actually know what those specific setbacks are, how they build, or what Kenisha does as a result. You might not even know what Kenisha's job *is*. A cursory glance at your Idea List will no doubt reveal many plot points that made total sense when stated as a simple, declarative sentence, but that once you ask yourself, *Um, so how exactly would that actually work?* become uncomfortably vague. It will feel a little intimidating, not because there's something wrong with you and if you were a real writer you'd automatically be filling in these blanks without breaking a sweat, but because the process itself is intimidating, for experienced and fledgling writers in equal measure. Facing that fear is part of what makes you a pro. Not to mention, flat-out brave.

Jennie had to face this kind of big, scary blank more than once, starting at the very beginning. She knew that Ruby was going to snatch a famous dog, who would instantly become the subject of a massive dog hunt, and that she would retreat to a hotel to try to write her script. Right now that's the bulk of the plot of her novel. That's what Jennie saw. But wow, how insanely vague is that? How could she actually *write* that story? As Gertrude Stein would have taken great relish in pointing out, "There's no there, there." Looking at the empty swaths in your novel, you might be thinking the same thing. It's enough to make many a writer panic. Don't. Instead, maybe take a nice walk around the block, clear your head, take a deep breath, and remember, although a *finished* novel is made up of several layers so deftly woven together that it feels all of a piece, when you're writing it, it's impossible to focus on all the layers at once. Meaning that once again, you're exactly where you should be.

Here's a handy reminder of what you're aiming for: the road between the gaps in your blueprint must be conflict laden. If it's an easy trek for your protagonist, if it doesn't spark the third rail, it's going to bore the reader. That's why everything your protagonist does to make the situation better should only make it worse, and the problem that much harder. And *not* just externally—as in, first your protagonist has to lift a 5-pound barbell, then a 10-, all the way up to 350 pounds. Sure, it gets harder, but really, who cares? Unless, that is, each external twist makes the internal struggle more acute, accelerating the battle between his desire and his misbelief. Never forget that your protagonist wants *two* things, which he's about to find out are mutually exclusive: to achieve his desire *and* to remain true to his misbelief. This is not because he's greedy, or a dolt. It's because so far his life has assured him that this is possible, and now he's smacking into the realization that it's not. So he struggles, confused, and often makes the wrong decision. Not to mention the least "risky" decision, in the hope that he can find a way to finagle both. Or, at the very least, get what he wants while giving up the least amount in return. And how often does *that* work?

The point is this: Always make it harder for your protagonist. Never give him the benefit of the doubt. If a bad thing could happen, let it happen. In fact, make it worse than he imagined it could possibly be—worse than *you* imagined it could be at first blush. Not to put too fine a point on it, but never forget that it's your sworn duty to continually yank the rug out from under him, to see how deep he'll have to dig to tap into the as-yet-unimagined inner resources he'll need in order to stand up and keep going.

With that in mind, let's see how Jennie might be faring as she tries to chart the external cause-and-effect trajectory between the moment Ruby decides to snatch a dog and when she's holed up in a hotel, five minutes away from the world discovering that she has Rufus. We already know why she nabs Rufus, but what we don't yet know is, *Why,* exactly, does Ruby flee at the moment she does?

The good news is that very often the answer to "what happens next?" stems from the answer to the last question you asked. That's exactly what happened to Jennie, who'd just nailed why Nora had the power to oust Ruby. Jennie's goal is to nail the third scene in her novel, the one that sets Ruby's odyssey in motion. She knew that she wanted Ruby to hit the road almost immediately, so she needed a reason for her to run built into the next scene in which Nora returns, ready to haul Ruby to Ojai. The question was, Even if Nora recognizes Rufus herself, why would that make Ruby flee? Here she is working it out:

> I need someone besides Nora to show up who might recognize Rufus. I'll give him a signature yap—a loud, easily identifiable bark, so even if he's locked in the next room, it would give him away. Okay, so what if Nora is so serious about getting Ruby out of the house and under her watchful eye that she actually engages a realtor to come to the house. Maybe Nora intends to rent the house out so Ruby *can't* stay. That's her big power play. It's kind of badass of Nora, but I like that. It fits. So this realtor shows up right after Nora recognizes the dog, and Nora tells Ruby to get out of the house and get the dog back to its rightful owner lest the realtor hear or see the dog.

And that is how Jennie figured out what happens next: by going over what she already knew about Ruby's story and searching for the logical event, *given who her characters are,* that would trigger Ruby's flight. Ruby is now officially on the run.

What next? As we've seen from the start, it would be dazzlingly easy for the plot to now revolve around Ruby deftly dodging dog hunters at every turn. Which would mean Jennie's novel would descend into one long, static chase scene. Not what either Jennie—or the reader—wants. That said, avoiding anyone who might recognize Rufus (which in the world of Jennie's novel is just about everyone) *is* what Ruby will have to do from here on out. Which means one thing: Each time Ruby dodges someone, there must be an internal component, some way the interaction triggers her internal struggle.

Ruby's first stop was easy. Jennie knew what Ruby would do once the rug was pulled out from under her: go to Henry, no matter what shape he was in. He's been her touchstone for years. So, again *logically, given who Ruby is*, Jennie knows that without thinking about the dog or the risk of its being seen, Ruby would instantly head for the hospital. Bingo! That's Scene #4, and a new card started. Even better, the next scene suddenly became a little clearer, because knowing that in a story things go from bad to worse, Jennie also knows that when Ruby gets to the hospital, things will not go well.

WHAT TO DO

Now you try it. First, take some time to gather up any empty swaths of your fledgling cause-and-effect trajectory that are now bereft of concrete plot points—basically, anything you've summed up in general. Your goal is to start bringing them into focus and making them specific. No matter how big the swath, all you need is a single specific "What" to begin digging for the "Why." Never forget: specifics lead to other specifics, and the light that illuminates the connection between them is always, "Why?"

Plot Size Matters

Having discovered the "Why" behind much of what will happen in your novel, chances are you've started envisioning scenes and your plot is beginning to take shape in your mind. As a result, you probably have a notion of the likely scope of your novel, be it a saga that spans three generations or a thriller that unfolds from dusk to dawn. This is a very good thing, because that knowledge will help you focus on the sort of external events that your protagonist will have in store. Take a moment now and think about the time frame your novel will span—do you see it playing out over years, months, days?

Here's Jennie thinking about her novel:

> I always pictured this novel unfolding over a short period of time and on a small stage. I *love* stories like that because they are so interior. Everything usually turns on something small. Which means I will get Ruby to the hotel she'll hide out in pretty quickly. Especially since I also want Ruby to be largely alone with the dog—I see Ruby and Rufus having the same problem: they're Internet famous, but their "public persona" is based on other people's versions of who they are.
>
> For Ruby, shunning social media is a continuation of her lifelong aversion to putting herself out there, allowing her real self to be seen and vulnerable. She didn't realize that as a result she was inadvertently allowing other people to define her. Now she wants her image back. And so, she discovers, does Rufus. No, he won't become a talking dog. But . . . I'm thinking that as a result of his "fame" he's become a neurotic, anxious, high-maintenance dog.

A neurotic, anxious, high-maintenance dog Ruby has to *hide* so she can write the script. This is a great example of the symbiosis between the deeper, inner story and the external plot that fuels it. And it led to yet another question: *Why* is Rufus so jumpy?

I wouldn't have blamed Jennie if she'd put her hands on her hips and said, "He's a *dog*. How the hell would I know! Now you want me to get into *dog* psychology?" Um, yeah. But she didn't say that, because she was already a step ahead of me. And notice how the more you know about the specifics of your story, the easier it gets to answer these questions. I can't say it too loudly (or too often): specifics beget specifics.

Here's Jennie riffing on what it is that gets to Rufus (and yes, even dogs can have backstory):

> I'm thinking that Rufus was a pretty laid-back dog to begin with, but he didn't like photo shoots and was getting increasingly skittish around

cameras, crowds, and bright lights. That wouldn't be good for Tony, either. Maybe ad execs were pissed off and coming down hard on him to get his dog under control. Maybe it got so bad that whenever Rufus heard the click of a lamp he began to howl and refused to eat. And so now with Ruby, all these neuroses are playing out—maybe Rufus is barfing all the time like Ruby was, and barking all night and not sleeping. The hotel is not happy—guests are complaining. And Ruby is not writing.

Great! This info goes right into Rufus's character bio. But Jennie needs to be careful, because as you can see, once again the dog plot is tugging at the story's leash. This is something you will run into throughout—because your plot is made up of tangible, visible events, they can have a near-gravitational pull, luring your story up to the surface and keeping it there.

To keep her plot from yanking her novel off point and dragging it into the middle of a big empty field, Jennie knew she had to begin to concentrate on what, exactly, would happen in relation to Ruby's *inner struggle*. That meant developing surface events that revolve around the script she needs to revise, Clementine's campaign to rewrite the ending herself, Nora's struggle with how to best help Ruby, and a lot of the things Ruby will be thinking about when she's locked in that hotel room. In other words, in order to move the plot forward, Jennie needs to begin to layer in subplots and secondary characters.

This is why it will take you awhile to fully develop your first five Scene Cards. Because—just as in life—in a novel everything is intertwined, so what looks like a single step forward is actually several steps, all taken in unison. As physicist M. Mitchell Waldrop said of the universe, "Everything affects everything else, and you have to understand that whole web of connections."[1] The same is true of the world of your novel.

We're going to begin to weave that web in the next chapter, as you start to develop, and layer in, the subplots and secondary characters that will give your novel the kind of depth that entices readers to step inside and leave the real world behind. But first, to better weave that web, let's figure out just how much territory it will cover.

WHAT TO DO

Take a little time and sketch out the scope of your story the way Jennie just did. What is the time frame, and how big is your stage? Consider the external plot and the internal struggle. Keep it short, always peering beneath each "What" to find its underlying "Why." If any new "Why" questions arise—and chances are they will—take a bit of time and pin down the specific answer, just as Jennie did with the source of Rufus's angst. It'll save you a lot of angst later.

14

THE SECRET TO LAYERING: SUBPLOTS, STORYLINES, AND SECONDARY CHARACTERS

In a good play, everyone is in the right.
—FRIEDRICH HEBBEL

There's no denying that by now some of you may have a pretty good idea of what your main storyline will be. In fact, given the work you just did, it may all fit together in what already seems to be a crackerjack cause-and-effect trajectory. Sure, the road ahead might be a little foggy, the twists and turns indistinct, but you know where you're headed, so why not just plow forward, get that first draft down, and *then* worry about all these other layers? Who knows, maybe you won't even need any other layers.

Do not go there. Because, as heartbreaking as this might be, if you *can* see your cause-and-effect trajectory that clearly at this stage, chances are your novel will be kind of dull. So before you put pedal to the metal, there's one thing you might want to consider first: the danger of tunnel vision. Even if you can see the road ahead of you, it's *all* you can see, and that's surprisingly treacherous.

The problem with solely focusing on the main storyline as you develop your story is that it soon becomes predictable and surprisingly flat, because such a narrow outlook flies in the face of what we were just discussing:

everything that happens affects, and is affected by, everything else. We are all connected, and very often those connections are precisely what reveal the "Why" behind what's *really* going on.

Okay, but clearly you can't develop *everything* that affects your protagonist, and by extension, *everything* her actions then set into motion. So what part of "everything" are you looking for? That's where subplots, storylines, and secondary characters come in, giving your novel the depth, scope, and dimension needed to magically evoke the sensation of real life for your reader.

Subplots often unfold on the same page, and in the same sentence, as the main storyline. That's what makes it tricky when you're envisioning your novel. In a finished novel, every layer has already been deftly woven in, so it's natural to assume that's how a story unfolds when you're *writing* a novel, as if every layer emerges simultaneously, already entwined with the others. Not so.

The layers of a novel are built exactly like the layers in recorded music. When you hear a finished song, everything melds into a single, transcendent experience. But that's not how it was recorded. Back in the studio, each track was created separately, beginning with the drums, the baseline, the foundation, and then each subsequent track was created so that when it was laid in it melded with the others. And the last track? The vocals, the voice, the thing we remember most is, as it turns out, the last track to be added, only after all the tracks beneath it have been expertly mixed into the seamless background that allows that voice to soar.

When blueprinting a novel, each layer is laid down one by one, so while *in the end* every scene will advance multiple subplots, deepen characters, and foreshadow the future, each of those layers was developed and woven in separately. In other words, what looks like one fluid step to the reader, for the writer is actually several individual steps taken in unison, synchronized into what *appears* to be a single movement.

To achieve that kind of seamless experience for the reader, in this chapter we'll explore how to find and develop the subplots already taking shape in your blueprint; the secret of creating secondary characters who will serve your story *and* their own agenda; and how to develop relevant backstory for each character so *you* can see the world from their point of view in each scene.

Subplots: More Than Meets the Eye

As you develop your novel's subplots, it helps to keep in mind that readers tacitly assume that everything in a novel is there on a strict need-to-know basis. If they didn't need to know it, why would you waste time telling them about it? That's why if you tell them something that does not affect the story, chances are at first readers will bear with you, figuring that its significance will soon be revealed. Until then, they eagerly second-guess you, assigning it story significance on their own—significance that is, by definition, wrong, because in truth, there is none. This is especially true of subplots because they take more page time than a single, errant digression, so it's important to remember that you must create each subplot with one question in mind: How will it affect the main storyline? Which, of course, has itself been created with one question in mind: How will it affect the protagonist's struggle? If you're writing a mystery or crime novel, this remains true as well: on the surface everything revolves around solving the crime, and the third rail is driven by how the protagonist makes sense of what's happening *and* why the perpetrator did what he or she did.

Every subplot must spin off the main storyline, telling the reader something they need to know if the plot itself is going to make sense. For instance, imagine *To Kill a Mockingbird* without the unlikely Boo Radley to rescue, and be rescued by, Scout. Imagine *The Great Gatsby* without Tom and Myrtle's tryst to trigger Gatsby's ultimate downfall. Imagine *Pride and Prejudice* without that dastardly fortune hunter Wickham and the impetuous Lydia, the better to unite Mr. Darcy and Elizabeth. You can't, can you?

Without those subplots there would *be* no story. Because subplots are, in fact, central to the story itself. That is, provided they don't take on a life of their own and start madly dancing to their own drummer. This happens more frequently than you might think, even to experienced writers. Subplots can be seductive, because there's no denying that many of them could tell a very captivating tale on their own. Who knows, if Jane Austen hadn't had her tongue tucked inside her cheek and meant *Pride and Prejudice ironically,* the story might have belonged to Wickham and Lydia, with Darcy and Elizabeth as the stodgy old spoilsports out to ruin their fun. The problem is, when subplots take off on their own, leaving the main storyline behind, they don't just abandon it; they effectively stop it in its tracks. So how do you search for subplots that will instead move your story forward? By taking to heart that otherwise rickety lesson Dorothy learned upon her return from Oz: when it comes to subplots, at least, you don't need to look farther than your novel's own backyard.

Where to Dig for Subplots

The term "subplots" tells us exactly where to look for them: beneath the plot, or, more accurately, beneath the surface of your story. Not beneath as in the random assortment of things that seem to collect under the bed all on their own, but beneath as in an integral story layer that, once exposed, sheds light on the surface meaning. It's not surprising, then, that a novel's principal subplots tend to spring from two story areas that you've already developed and that often overlap:

- External events that were set into motion before the novel began, and that have impending consequences that will affect the protagonist's quest
- Secondary characters (basically, anyone other than the protagonist)
  Much like drilling down into your plot points, creating subplots will send you back into the past to dig for the specifics that will make them relevant to the story you're telling. Your goal at this stage is not to develop the entire subplot from start to finish, in intricate detail. Rather, the goal is to begin to envision it, where it might go, and how it might spur your main storyline. When you're first crafting subplots, you'll not only create

new scenes; you'll also expand existing scenes, which you'll note on the "Subplot" line just beneath the Alpha Point on the appropriate Scene Card.

Here's the secret: subplots tend to revolve around one character. For instance, in Jennie's novel, Nora, Clementine the fangirl, and Sharon the producer will each have a separate subplot. Henry, too. So when you're noting the subplot's main point on your Scene Card, it's often easier to focus on *whose* subplot it is, rather than its topic. Why? Because when it comes to envisioning a scene, it's far more concrete and so much easier to imagine—and feel—what a character is doing, than to focus on what, conceptually, will happen in a particular subplot. So in Ruby's story it wouldn't be "script subplot" but "Clementine's subplot" or "Sharon's subplot," depending on who is in the scene.

The reason these cards have several spaces for subplot info is because each scene will most likely move several individual subplots forward in one fell swoop, and the more aware you are of each one, the better.

To help you get the hang of it, let's watch Jennie as she starts developing one of each type of subplot, and then either add it to her Idea List, begin a card for it, or, if it's part of a scene she's already envisioned, add it to an existing card.

External Event Subplots

Subplots that are already in motion when your novel begins are powerful because they hit the ground running, giving the protagonist no choice but to take action. This is something we've been talking about from the get-go—your story begins in medias res, and the same is true of many of your subplots. They've been picking up steam since long before your novel began. Take a minute to think about your story, look over all your existing plot points, and see if you can pinpoint the balls that are already in play.

Jennie went back through her Idea List and Scene Cards and identified several possible subplots:

- Word got out that the studio halted production and called for a rewrite, and fans are speculating on why, what the original ending was, what Henry's absence will mean, and whether Ruby can pull it off on her own.

- Sharon, the producer, has a lot riding on the end of the show. She might even be pissed that Henry decided to end it, and certainly nervous about getting her next gig. She is no doubt hounding Ruby to come through.
- Clementine, the fangirl who has been picked to revise the show's ending if Ruby doesn't produce, definitely has a subplot that's already in play.
- Henry will never awaken, but he will have a major subplot in the form of flashbacks as Ruby sifts through and reevaluates their relationship—after all, he's the reason for her debilitating grief.
- Ruby has been barred from the hospital—not allowed in to see Henry, maybe because she's not technically family?

It's no surprise that at this stage what Jennie knows about these subplots is, well, next to nothing. Take the last one: *Ruby has been barred from the hospital.* This is one of those unassuming declarative sentences that at first glance certainly seems potent, but when you really focus on it, you're left with a lot of questions. Like who's barring her? And why? And how does it affect what's going on with the dog and the script? But it's an excellent start, because it gives us a very concrete specific that we can begin to explore and expand. That's the ball in play that Jennie decided to develop.

WHAT TO DO

Look over your Idea List and your Scene Cards and do what Jennie just did—identify anything that you think might be a potential subplot. Make a list of the possible candidates in order of appearance in your novel. Take the most interesting one on the list and start tossing it around.

Here are Jennie and I doing just that:

LISA: So why can't Ruby see Henry in the hospital?

JENNIE: Because she never married Henry. Only family can get into ICU.

LISA: I'm not sure that's technically correct. Anyone can visit—unless there's a restraining order.

JENNIE: Okay, so what if someone has made it clear they are simply not welcome? That could be Henry's mom; let's call her Frances. She could not want Ruby there.

LISA: That's good—but now you have to ask why Frances would try to shield Henry from Ruby. What's her relationship with her son? And with Ruby?

JENNIE: Henry is her only child, she loves him dearly. So she would obviously not move from his bedside.

LISA: True, but that's a given. We'd expect any mother to do that. It doesn't tell us anything about the dynamic of their relationship.

JENNIE: Okay, so maybe Frances lost her husband, or even better, maybe he abandoned her when Henry was a little boy and she poured all her love into Henry. Maybe Henry felt responsible for his mom's happiness, and it led to guilt and angst—he'd probably felt pressured all his life to "be there" for her, so that a woman like Ruby was a major relief.

LISA: Which is why Frances would be suspicious of her. And around the time of the wedding? She'd be on high alert.

JENNIE: True! If she'd been abandoned, then when Ruby seemingly "jilted" Henry, she'd be doubly fried. And maybe she'd try to protect Henry even more fiercely.

LISA: Which might make Henry feel stifled, smothered, and give him empathy for Ruby's fear of marriage. Ironically, it might make him love Ruby even more.

JENNIE: Exactly! And you really can't blame his mom for how she feels either.

LISA: That's the point; when you understand where people come from, and why they're doing what they're doing, you often realize that what looks like "bad" behavior, isn't. With very few exceptions, everyone believes they're doing the right thing for the right reason.

And there you have it: yet another example of the way specifics do indeed beget specifics. Now we know, specifically, how and why Frances's distress will impact Ruby, and why it will matter. Did you notice, too, that when Jennie began exploring Frances's storyline, it dovetailed with another ongoing subplot—Henry's relationship with Ruby—giving us insight into why Henry may have loved Ruby even *more* because of her independence? This is a perfect example of how the individual layers of story intertwine to create a web that, on the surface, looks solid and all of a piece.

And better yet, here are five possible scenes that emerged from Jennie's brainstorming session:

- There will be a scene at the hospital when Frances and Ruby first see each other after the accident and all their past resentments surface.
- There could be a flashback scene to the day Ruby and Frances first met—with Frances's expectations for Henry clashing with who Ruby really is.
- There would definitely be scenes with Frances on the day the wedding was canceled.
- I'm thinking we'll get the story of how Henry's parents broke up—perhaps from Henry's point of view in a flashback, or maybe in a scene between Ruby and Frances at the hospital. Or maybe both, and Ruby will have a small "aha!" moment seeing the discrepancies in their stories, and realizing that Henry saw the breakup of his parents' marriage very differently than Frances did, and so they spent decades painfully misreading each other.
- And, a final scene at the end, when Ruby comes to Henry's hospital room just before he dies (I'm thinking he'll be on a ventilator, and this is when they've decided to turn it off), and Ruby and Frances have their final confrontation.

Will all these scenes be in Jennie's novel? Maybe. Maybe not. But since each one stems from Ruby's story-specific past, and each one is relevant to what will happen in the plot, Jennie has some very potent possibilities to plumb here.

So she quickly began a Scene Card for each possible scene and put it into her Development Folder. She also started a character bio for Frances.

WHAT TO DO

Take a look at the list you just made of possible subplots already in play when your novel begins. The good news is that if it's on your list, you already know it has story relevance. The first question to ask is always, Why will this matter to my protagonist, given her quest? The answer, even when it's fuzzy, will tell you where to begin digging into the past, so you can zero in on the precise events that will make it specific, concrete, and viable. Chances are you'll end up with several potential scenes for each subplot you've identified.

Secondary Characters and the Subplots They Ride In On

While many of the characters in your novel will likely spring from your protagonist's past, some—like Tony, the dog owner—will not. As your plot unfolds, there will be characters your protagonist won't have met until you throw them together right there on the page. So who the heck are they, and how do you keep them from becoming puppets who simply do what the plot needs them to do, rather than what the *story* needs them to? Or, for that matter, what *they* would do themselves? After all, every character has their own agenda, and a very subjective lens through which they see the world—which, you can bet, is going to be very different from your protagonist's. Once you've created these individual characters, how can you make sure that in addition to staying true to themselves, they also remain true to your novel? The trick is to allow them to dance to their own drummer, as long as that drummer is playing your novel's tune.

What They Don't Know Won't Hurt Them

Think about your own life for a minute. In it, you are the protagonist—it's your life *story* after all—and while you dearly love your friends, parents, significant other, and, unlike Ruby, Fido, let's face it: they're supporting players. Critically important players, players you'd be lost without, but at the end of the day, well, it's not really their story, is it? Of course you'd never mention that to them; why hurt their feelings? But here's the thing: in their lives, *you* are a supporting player. The point is, every single character in your novel fervently believes he or she is the protagonist with the same unquestioned assuredness that we all do, and as such, everyone else is there to facilitate *their* agenda, not yours, your plot's, or your protagonist's (or so they think). And that's as it should be. Everyone everywhere is clever in their own way and has their own agenda, and everything they do is calibrated to achieve it. That's biology in the guise of human nature, and a character without a clear, subjective agenda is neither believable nor engaging.

What does this mean for you as a writer?

It means when you develop your secondary characters you need to think about them in the same way you think about your protagonist. Each character has his or her own driving agenda, realizations, and, often, their own arc.

If you're not aware of what your secondary character's agenda is, it's easy to forget they have one at all. Thus you inadvertently end up focusing solely on making sure that their action conforms to what the plot needs them to do in each scene, which means that over the course of your novel their actions won't add up, because there's no internal logic guiding it. To be very clear, we're not talking about a shifting daily agenda—one that's different for each scene, each day, each hurdle—but a single, overarching agenda that drives each character's behavior throughout the novel.

Here's the secret. Since you know that none of these characters is the protagonist, you are going to create them, *and their agenda,* with one purpose in mind: to help facilitate the protagonist's story. In other words, unbeknownst to them, they *are* there first and foremost to serve your protagonist's struggle, and so everything they do is geared to have a specific effect on your protagonist, sparking your novel's third rail. This means that although each one of them could stand alone as a full-fledged human being, and everything they do in their lives will make sense given who they are, you'll create them and their beliefs precisely so they will *naturally facilitate your protagonist's story.*

Does that mean you need to dig as deep into your secondary characters as you did with your protagonist? No. But you'll need to dig deeper into some characters than others, depending on the role they play. For instance, if you're writing a love story, then you'll probably need to know just as much about your protagonist's significant other as you do about her. But those losers she dated along the way? Not so much.

Take a minute and make a list of all the characters who have appeared so far, then select those who you know will play the most significant roles in your novel. For instance, in *The Godfather* a good start would have been Vito, Sonny, Fredo, and Kay, but not Joseph Zaluchi. If you're wondering, *Joseph Zaluchi, who's he?*—point made.

Jennie's already begun character bios for Nora and Frances, and that's exactly what you're going to do for every character who will play a significant role in your novel. These are not bios of the "Nora, *this* is your life!" variety. Instead, as with the bio you did of your protagonist, it's story specific. The most important thing to keep in mind as you develop your secondary characters is that each one must be conceived and developed in accordance with the role he or she will play in your protagonist's story. Thus the first question to ask of each one is *Okay, how will this character's agenda fit into the story I'm going to tell?*

One last thing before we take a stab at developing a character. If you're writing a mystery, a thriller, anything at all involving intrigue, you *must* know your bad guy's agenda, and—this is utterly crucial—what he or she has done thus far to actualize it. For instance, if they're a murderer, and your novel is about how your detective catches them, you need to know, in specific detail, whom they killed, how, why, and where *all* the clues are buried *before* you send your detective into the field to hunt for them. Without that knowledge, how would you, let alone your detective, know what to look for, where to search for it, or the significance of anything they found? The answer is simple: neither of you would have a clue. This is another perfect example of the importance of knowing the *first* half of the story, so that your novel can begin on the first page of the second half.

I asked Jennie which character's subplot—also known as their storyline—she wanted to tackle first. She chose to work on Tony, the dog owner, because she liked what emerged for that character when she wrote the "aha!" scene and felt that it could be central to her story.

WHAT TO DO

Now it's your turn to pick a character to develop. If you're writing a romance, your protagonist's beloved is a great place to start. If you're writing a mystery or a thriller, then your best candidate is the bad guy (or spy, or political operative—in essence, your protagonist's nemesis). If, like Jennie, neither of these apply, look over your list. Is there a character who's madly waving his or her hand in the air, screaming, *Choose me, choose me*? If so, go for it. Otherwise, start with the character who will appear first in your novel.

Remembering that your goal is to *create a character* who'll either challenge or reaffirm your protagonist's misbelief, ask yourself, What, in general, might this character open my protagonist's eyes to? Write it down, even if it's just a wisp of an idea, even if it feels fleeting and wobbly, or cliché.

Here's what Jennie wrote:

> Tony will show Ruby what true love looks like. I know that sounds insanely stupid, but that's it. Tony is willing to be 100 percent vulnerable, to risk turning the whole world against him, to stake this very public claim for this girl whom he hopes will love him back. Ruby sees—right in front of her—what she failed to do with Henry, and she will realize that she still has one last opportunity to make good on her love for Henry, by what she does in their script.

Great. The next step is to develop Tony's backstory, so he will be in a position to do just that. I asked Jennie to show us her work in fleshing out Tony. I expected her to take her time, first combing through what she already knows about Tony, then painstakingly writing draft after draft until she felt she nailed him. Instead, she came back immediately—we're talking within minutes—with something that was fully formed. Let's take a look at what she did, explore why it's so good, and then take a guess at why she was able to write it in record time.

Tony, the Dog Owner

- Okay, so this dude got his dog because he thought it would help him get girls.
- I am seeing him about twenty-seven years old, handsome, maybe a gym rat. He's a weightlifter guy. A trainer. But socially inept—thus the need for the dog.
- He picked this particular dog from the pound because some girls told him rescuing a dog is cool. At the pound a cute girl was all into this ugly dog and thought Tony was great for even considering it. So he got the ugly dog.
- He started posting dog pictures on Facebook and Instagram and on his Tinder profile to get girls. But the *dog* started getting all this attention and all these followers, and it gets out of hand and suddenly the dog is getting endorsement deals and media gigs and Tony has to cut back at the gym to manage it all.
- He digs it because he gets some girls.
- Then it gets ridiculously out of hand—travel, appearances, constantly catering to the dog. And the dog is starting to get angsty—barking all the time, skittish in crowds, barfing when he's on camera. As a result, Tony can't actually keep a girl because he's all about the dog and he's gone all the time. And besides, he doesn't know how to connect, and he's scared of it anyway (like Ruby) . . . but now, as opposed to before, he really wants to connect because he is jealous of the love people have for his dog.
- When the girl he really does like leaves him (because she figured out that his love for the dog was BS, and she then assumed that his love

for *her* was BS), he realizes she was right about Rufus, and so decides the dog has to go—right now.
- So he heads to the dog park, thinking of coyotes—and happens to let the dog go five minutes before Ruby drives up and snags it.
- The Internet goes insane with people searching for the dog, forcing Tony to act the bereaved dog owner.
- At the end of the story, Tony and Ruby will both come to a realization together about dogs and people and love and loss.

Why This Character Bio Works:
- It's a complete cause-and-effect trajectory—each event triggers the next.
- It's specific; every bullet point is something that actually happened.
- It tells how each event affects Tony internally, and the meaning he reads into it. Thus we know why he makes every decision, takes every action, and what his underlying motivation is at each turn.
- It addresses only the parts of Tony's life that are relevant to Ruby's journey. It covers the dog, Tony's desire for love, but *not* his no doubt tragic, but irrelevant, painfully awkward high school years.

This is one of the best initial character bios I've read. I asked Jennie why she thought she could come up with it so easily, when to do the same for Ruby was like pulling teeth. "Well, for one thing," she said, "I had that magazine article about the ugly dog for inspiration—an actual person whose life had been turned upside down for their dog. And because of the work we did with Ruby's misbelief, I knew one key emotional thing—which is that this dog owner was *not* going to be heartbroken that his dog was gone. From there it was pretty easy to imagine it all. Having constraints and limits really helps—which is why it gets easier and easier to write a story as you get into it."

But there's a bit more to it than that. This is a phenomenon I've noticed in countless writers as they dive deeper and deeper into their stories. It doesn't happen overnight, but when it does the results are spectacular. You'd think they were channeling the muse, but what's really going on is even more amazing, because it's something you can control, and therefore count on. It's this: once you've created your story, and your characters have taken on the shape, depth, and complexity of a fully formed past, they begin reacting to the present as if on their own. The same is true of events, subplots, everything—they begin to suggest themselves, based on what has happened up to that moment. Because Jennie knew what Tony's role was in her novel, she was able to create a character who wouldn't be *playing* that part, but who would be *living* it. Soon, chances are you'll find the same thing happening to you.

WHAT TO DO

Create a bio for every secondary character you've identified thus far, just as Jennie did for Tony. Take a look at each one and see if there is anything that leaps out at you as scene potential. For instance, when Jennie reviewed Tony's bona fides, one instantly leaped out at her: the Internet goes insane with people searching for the dog, forcing Tony to act the bereaved dog owner. Here's what Jennie wrote:

> I can imagine Ruby hiding in some hotel, watching TV, and seeing a clip of Tony, *seeming to* bare his grieving heart for all the world to see. And maybe Ruby is thinking: *I don't buy it.* After all, she's spent a good chunk of time around actors—pros and fledgling actors alike—and can probably tell that this guy isn't grieving as much as he'd like people to believe. And *she's* so deep in real grief that fake grief would totally rub her the wrong way. Ruby becomes suspicious of his motives and starts digging to learn more about him and Rufus.

So what did Jennie do? She instantly started a Scene Card, whose Alpha Point is "Ruby realizes Tony isn't as upset about Rufus as he claims to be" and slipped it into her Development Folder. Notice that this fact will no doubt affect the "aha!" moment scene Jennie has already written, but rather than revising it now, she simply made note of it on the scene itself.

The point is, the more you know your characters, the more your novel begins to write, and revise, itself.

Once you know what brings each character to your novel, the question becomes, what will they do when they get there? Each one has his or her own subplot. The more story-specific info you uncover about them, the clearer their purpose will become. Your end goal is to be able to see through their eyes as well as you can see through your protagonist's. Because before you can write any scene they're in, you must first put yourself in their shoes, so you know exactly what their agenda is, and why.

Gone But Not Forgotten—Creating Characters Through Backstory

But what about characters who'll play a large role in your novel, and yet might never actually appear in it? I'm talking about characters the reader gets to know through backstory, flashbacks, and other characters' memories of her or him, but who don't have a role in the present-day story. Yes, I'm talking about Henry. You might have such a character in your novel, one who is so entangled in your protagonist's third rail that even if he never appears, or doesn't appear until the end, he will be as vivid, alive, and conflicted as everyone else.

Here's the skinny: these characters have to be as fully developed as any other character, and it pays to develop them early on. I didn't want to push Jennie too hard, because she had her heart set on getting Tony, the dog owner, onto the page, and clearly she was right to work on him first, given how well his bio turned out. But now, before she goes forward, it's crucial that she spend time developing Henry and his storyline—a storyline that, as far as poor Henry is concerned, will have pretty much ended before the novel begins, but—and this is the point—will play a huge part in driving Ruby's storyline.

So with that in mind, Jennie sat down to flesh Henry out. She kept in mind that Henry had to be his own man, and at the same time, that man had to be someone who would best serve Ruby's story. Here is her first, rudimentary take on Henry:

- Henry is three years older than Ruby (who I've decided will be forty-seven. I want her to be safely past the age when she could have children, but still have a lot of life left. This also works with the real-life timeline of my story, which has to incorporate the reality of Web-based TV, which heated up in the early 2000s.)

- Son of a prominent lawyer-type dad, Henry Sr., and society mom, Frances—an old-fashioned marriage arrangement. Their marriage blows up when Henry Sr. has an affair, leaving Frances high and dry (I can have betrayal in a secondary character—so fine with that!). Frances would be adrift and devastated and pissed. Henry would have a deep aversion to that life—of being the man on whom a woman is so dependent. Which is why he has never gotten serious with a woman. He couldn't be "the provider" to anyone, couldn't stand that whole settling-down thing. So a woman like Ruby who doesn't want the traditional marriage gig would appeal to him.

- He wanted to be out in the world, his own boss, not a company man. (Oh! Henry Sr. could have an affair with a young associate at his firm—which would make the corporate life that much more repugnant for Henry.) He wanted to *create* things but never knew what exactly, or how.

- Grew up somewhere, not L.A.—Texas? CT?

- Came to L.A. and studied English to rebel, which is where he'll meet Ruby.

- Has done a ton of jobs, bounced around—got job as a journalist but didn't love it, because he wasn't his own boss. Worked for a while as a cook? On a fishing boat? Worked for a guy who shaped surfboards. Was handy with tools. Was hired to work on a movie set. Got into that, and . . .

- Always thought he could come up with something better, storywise, than the movies and shows he worked on, but never dared. Was chicken to *really* rebel. Until he met Ruby . . .

This is a great opening bio, but as you can see, it's not as detailed, or quite as attuned to Ruby's story as Tony's was. At this stage, that's fine. Don't forget, because Jennie hasn't fully fleshed out Ruby yet (she only just decided how old she'd be), these two bios will probably evolve together. The more Jennie figures out about Ruby, the more she can layer in about Henry.

Why This Works:
- Did you notice that Henry doesn't want a traditional life, just as Ruby doesn't? This is good because it plays into what Ruby is struggling with.
- Did you notice that almost everything Jennie wrote was in some way connected with Ruby? Henry is coming to life to serve her story.
- Did you notice that Henry wants to create things, but didn't have the courage to do it on his own until he met Ruby? They need each other in order to become the characters we will meet on the page.

WHAT TO DO

Is there anyone in your story who, like Henry, played a big part in your protagonist's past? You're looking for characters who helped shape your protagonist's story-specific worldview, because they're who she will think of as she struggles to make sense of what's happening in the present. It might be your protagonist's mother, father, sibling(s), old flame, or even imaginary friend. Make a list of these characters, and take a moment to begin to envision them. Who are they? Write out a quick story-specific bio just as Jennie did with Henry.

That done, there's one last area to explore: How do you develop a character's backstory so that it will supply meaning in the present? Especially when chances are some of it will unfold *in* the present in the form of flashbacks?

Backstory *Is* the Story

The good news is that you may already have done some of this work in chapter 7 when you wrote your three story-specific backstory scenes, tracing the progression of your protagonist's misbelief from its inception up to the moment your novel begins. Because even if your protagonist has no shared history with anyone in your novel, she will bring her peeps with her in the form of memories, and those voices we all have in our heads who are all too eager to give us advice any time, day or night: Mom, Dad, BFF, and the mean neighbor two doors down who always suspected us of egging his house when we were twelve (okay, okay, that one was me).

Jennie's novel is a perfect example of how integral backstory is to what's happening in the moment. From the very beginning she said she was drawn to writing about regret. That's her goal. And according to Webster's, regret means *to feel sad or sorry about* (*something that you did or did not do*): *to have regrets about* (*something*). In other words, Ruby will be thinking about something in her past, reevaluating it, probing it for new meaning, and ultimately seeing it in a new light with fresh eyes. So the next logical question is, what things will she be thinking about? What, specifically, will she regret? We already know that the answer can't be something general like, "She regrets that she didn't marry Henry."

That means Jennie still has a lot of work to do. She needs to go back and concretize what, specifically, happened between Ruby and Henry. Not just so that snippets can run through Ruby's mind, but so the reader can experience it full-on, just as Ruby did back in the day. Yep, that means she'll be writing full-fledged scenes that reveal turning point moments in their past—some of them will be big events, like when Ruby and Henry call off their wedding at the last minute. Other scenes might be small moments that are just as profound, just as powerful—for instance, the first time they shared a joke just by glancing at each other and it felt like they'd read each other's minds. Or a moment when Ruby saw Henry do something vulnerable and found it so endearing that it left *her* feeling vulnerable, causing her to shut Henry out.

Jennie will be looking for moments—good and bad—when there was conflict, when Ruby and Henry didn't say what they were thinking, when they misread each other, when they were oblivious to what was really happening.

But in order to pinpoint those moments, she'll first be looking for the story-specific events that spawned them. For instance, when she wrote her opening scene she made an (anguished) mental note to figure out the history behind Ruby and Henry's show. Here she is riffing on the events themselves, in order to give herself a framework to begin to explore and build upon.

- In 1995 Ruby and Henry were working on a whiz-bang live action movie remake of the Rapunzel story— she as the costume designer and he as the grip. The script was dreadful, so in between takes they'd make up new dialogue and whisper it to each other, while trying not to laugh. In their version the girl did not need rescuing—and she climbed down out of the tower of her own accord.
- They got together for drinks and began writing on a napkin. Of course.
- They come together, but Ruby sabotages it, and they go their separate ways.
- In 2006 they ran into each other on another set. And they couldn't help it—they started rewriting the movie. It turned into a three-minute synopsis that skewered it.
- A friend of theirs was putting stuff up on the Web—totally unheard of at the time. One night, soon after the real movie came out, when a bunch of the actors were drunk, they shot the synopsis, the friend put it on YouTube—and overnight, it went viral.
- So Ruby and Henry did another, and soon they had a series of hits. Someone suggested that they do something longer and they decided to rewrite a classic play à la *Rosencrantz and Guildenstern Are Dead* and tell it from a totally different perspective and make it modern, edgy, funny—and episodic.
- It's a massive hit, and this becomes their job. They do it for seven years—reimagining classic plays as a modern-day TV series.
- When Henry gets the movie gig without Ruby and announces the upcoming season will be their last, they decide to tackle *Romeo and Juliet*. They disagree about how to end the show.

Jennie created that basic set of external events for one reason only: so she could then dig into it, searching out how it affected, and was affected by, Ruby and Henry's relationship. In other words, she forged a surface situation so she could probe what lay beneath it as it relates to the story she's telling.

And just to be clear, developing this kind of backstory doesn't solely apply to characters who won't appear in your novel. It applies to the shared story-specific history that every single character has with your protagonist.

WHAT TO DO

You now have story-specific bios for every character you've thus far identified. If any of them have a shared history with your protagonist, see if you can begin to flesh it out. You're not looking for everything they ever did together. For instance, Ruby and Henry might have spent every summer hang gliding off La Bajada Ridge in New Mexico, every New Year's Eve binge watching old episodes of *Father Knows Best* till dawn, and every Sunday afternoon square dancing, and it wouldn't matter an iota. (Oh all right, they never would have done any of that, because I just made it up.) But if they *had* done it behind Jennie's back, it still wouldn't be on the list. Because it has nothing to do with the story she's telling.

So keep your eyes on the prize as you make your list. You're looking for only the moments when what happened between the protagonist and the character in question has story relevance. Once complete, transfer anything still too conceptual or general to visualize to your Idea List, and begin Scene Cards for those that you can sit back, close your eyes, and actually begin to see. This time, you may very well be able to do that with most of them, with very few destined for your Random Folder. In fact, I wouldn't be surprised if a lot of your Scene Cards are nearing completion. Now what?

15

WRITING FORWARD:
STORIES GROW IN SPIRALS

It always seems impossible until it's done.
—ATTRIBUTED TO NELSON MANDELA

Let's take inventory. Right now you have character bios and backstory on all the secondary characters you've identified thus far, a fledgling draft of your first scene and your protagonist's "aha!" moment, and a slew of Scene Cards in various stages of completion in your Development Folder (not to mention several almost-rans languishing where they belong, in your Random Folder). A lot of your cards might contain nothing more than the first, general crack at their Alpha Point. That's fine. Your Idea List probably still has a few of those too-vague-to-envision ideas on it, waiting in the wings like anxious understudies hoping for their big break. Now it's time to pull out the Scene Cards you've been developing for Scenes #2 through #5 and bring them to completion, so you can start focusing on writing as well as blueprinting.

How to Play Your Cards Right and Get Your Story Straight

Your opening scene is written, and thanks to your tenacious digging, the cards for your next four scenes may already be fully populated, with your story logic layered and firmly in place. At this stage, if there are logic gaps they shouldn't be hard to identify, and with the skills you've acquired, you know what to dig for in order to fill them in. The trick is to make sure that each scene touches the third rail, forcing your protagonist to make a difficult decision that's part of an escalating cause-and-effect trajectory. You want to plunge us into action, and at the same time—this is the crucial part—give us enough insight into your protagonist to understand what the action means to her. In other words: We need to know what's *really* at stake.

Here's Jennie doing just that for her second card:

> OK, so after the opening scene in which Sharon lets Ruby know that if she doesn't deliver the rewrite on time, Clementine will, and Nora makes it clear she doesn't think Ruby is able to stay in her house alone, Ruby decides to "borrow" a dog to get Nora off her back. But the scene can't just be about the dognapping. It has to touch the third rail, and to do that we'd need to be in Ruby's head as she comes to this decision. (See, Lisa's voice is totally in *my* head . . .)

I hear that from writers a lot, actually, usually when they were about to take the easy way out, and then, ahem, didn't. So with that in mind, here's the thing Jennie needed to consider. While yes, we want to be in Ruby's head as she comes up with the dognapping scheme, it would be even better if something external happened to spark the notion. Because just as action is meaningless unless it's having an effect on someone, abstract thought is boring unless it's spurred by action. Plus, since Ruby's scheme about the dog really is magical thinking, perhaps it isn't the first solution she considers. Maybe something triggers the "aha!" moment about the dog, so I asked Jennie to brainstorm that.

> So the whole reason Ruby will think stealing a dog will placate Nora is because Nora has always wanted Ruby to get a dog and have a husband— which is to say that Nora didn't want her sister to be so damn above everything and everyone. Most of all, Nora wanted Ruby to need her. She

wanted to feel needed. The dog will not just be a random idea, but the one thing that Ruby is certain will convince her sister she is okay. I can show that realization coming over Ruby, and at the same time let the reader see that this is a woman who is really not thinking clearly . . .

Yes! Take a look at Jennie's second Scene Card on the following page. Did you notice how layered it is, right out of the starting gate? Everything that happens has a story-specific emotional component to give it meaning and drive it forward. On the plot line, we can see the arc of the scene, the external change that happens within it. That is, Ruby goes from trying to assuage Nora's concerns with a well-stocked refrigerator to "borrowing" a dog instead. This isn't merely surface logistics, as the third rail evidences: for Ruby it's proof of her ability to outsmart her sister.

And did you notice the *And so?* "Ruby speeds home, sure her problem is solved." Feel the implied conflict? It's clear that not only will Ruby's ploy not convince Nora that she's okay, but it'll prove to her that Ruby's just leaped off the high dive into a whole bucket of crazy. Talk about irony! Which means that Jennie knows exactly what has to happen next. Here she is working it out:

Nora shows up to find her sister has this dog and, worse, she recognizes the dog. It's a famous dog with a totally recognizable bark. Ruby is mortified. Moments later, the realtor Nora hired shows up—and Nora tells Ruby to take the dog and run, for fear the realtor will catch on. Ruby, now desperate to get away from her sister and her compounding problems, bolts. She grabs her computer, takes the dog, dashes out the back door, and dives into her car. I first imagined that she would go to the dog park and take the dog back, but that's too easy—and doesn't touch the third rail (does anyone see a pattern here?). So I decided that, panicked, Ruby would go where she always went when she felt beleaguered and alone— to Henry. So she goes to the hospital.

SCENE #2

ALPHA POINT: Ruby dognaps Rufus

TONY SUBPLOT: Unbeknownst to Ruby (or the reader) Tony watches as Ruby "steals" Rufus

	CAUSE	EFFECT
	What happens	**The consequence**
THE PLOT	• Ruby showers, puts on "real" clothes and heads to Whole Foods to buy "real" food to impress Nora. • At the market she sees a woman pampering a fluffy white "purse" dog. • It dawns on Ruby that everyone—Nora especially—says that having a dog provides emotional support.	• Ruby decides that a dog would definitely get Nora to back off, but she needs one in the next half hour, and doesn't want to actually keep it once Nora leaves. • Ruby decides to "borrow" a dog from the dog park for an hour or two. • Ruby nabs what appears to be an ugly mutt no one seems to be paying attention to.
	Why it matters	**The realization**
THE THIRD RAIL	• Ruby is desperate to keep Nora from forcing her out of her house. • The last thing Ruby wants to do is let her guard down with Nora. • But beneath Ruby's bravado, there is a longing that makes her uncomfortable. A small part of her wishes she could turn to Nora for comfort. A feeling she finds annoying. • Ruby hasn't noticed yet, but oddly, this is the most "present" she's been since before Henry's accident.	• Ruby believes she's outsmarted Nora once again. • Ruby is sure that, given how much the dog owner mindlessly doted on her fluffball, dog people are easy to please, and definitely not discerning or, let's face it, smart. **And so?** Ruby speeds home, sure her problem is solved.

Brilliant! It's not about the dog, it's about what the situation with the dog triggered in her. Loneliness, fear, panic. And so without thinking, she will instinctively head straight for the one person she's always gone to for comfort, but it won't be comforting at all.

Her third Scene Card on the following page is pretty straightforward; Scene Cards often are. Chances are when Jennie writes (and rewrites) the actual scene, new layers will emerge, as well as layers that she'll discover midway through the novel and then circle back and weave in here. And notice that the *And so?* always leads directly to the next scene. "Ruby goes to the hospital." If that made you think, *Wait, I thought Frances had banned Ruby from the hospital. And isn't Henry in a coma? Why would he be a comfort to her?* Good points. Here's Jennie working out the answers for what will become her fourth scene:

> My original idea was that Henry would be on his deathbed as the novel began—comatose and not going to recover. But now I realize that it would work better if he wasn't quite that far gone at the very start. So now, as the novel begins, there will be a chance he could come out of whatever has happened to him—I'll figure the medical part out later— which will give Ruby just enough hope to fuel her magical thinking: she'll get to the hospital and find Henry awake, she'll tell him the dog story, he'll get a big laugh, and they'll work out what to do with Sharon in terms of the show, and everything will be okay. Instead, Ruby will get there just as the doctors are telling Frances that there's no hope, he's going to die, and maybe that's when Frances actually bars her from even being in the room. So right when Ruby needs him most, he's going to be out of reach for good—

Did you notice that Ruby's story has begun to unfold based on what matters to her most, not on what she should do about Rufus?

SCENE #3

ALPHA POINT: Nora arrives and recognizes the dog

NORA SUBPLOT: Nora realizes Ruby is even more out of control than she's thought, but protects her anyway

		CAUSE	EFFECT
		What happens	**The consequence**
THE PLOT		• Nora arrives at Ruby's house, Ruby pretends that Rufus is her dog • Nora quickly recognizes Rufus—especially once he barks. • Nora goes online and proves to Ruby that he's a famous dog, and that there's already an intensive search going on for him. • Just then there's a knock at the door.	• Ruby tries to deny she dognapped Rufus. • Then she blames Nora for putting her in this spot—she's fine, and her plan was to take the dog back in an hour anyway. • The realtor Nora called arrives, and Nora's loyalty to Ruby trumps her concern about what she's done. • Nora tells Ruby to flee; she'll deal with the realtor, who hopefully didn't hear Rufus bark.
		Why it matters	**The realization**
THE THIRD RAIL		• Ruby was sure she'd solved her problem, and was gearing up to be a bit smug about it, chiding Nora for thinking she'd go to pieces like that. • Ruby's never allowed herself to be vulnerable to Nora; it's mortifying—instead of proving to Nora she's okay, she's exposed and feels foolish.	• Ruby realizes she's losing control of her own narrative and it terrifies her. • Realizing she has nowhere to go—no friends, no family, no safe haven—Ruby instinctively turns to the only person who ever made her feel safe: Henry. • Nothing else matters. **And so?** Ruby takes Rufus and her laptop and heads to the hospital.

In her fourth Scene Card on the following page, the *And so?*—that Ruby *can't* go home again without giving up entirely—is a big turning point for her. She's lost Henry, his mother has banished her from his hospital room, Nora is threatening to sell her house, Sharon is talking about letting Clementine write her show. Basically, she's toast. Except for one thing. She still has that damn dog in her car. Here's Jennie working that out:

> After the hospital, Ruby has to confront the problem of the dog—whom she totally forgot about. When she gets back to the car, several people are staring at Rufus, and Ruby realizes she could let the dog go. She could say she found him by the side of the road and was about to take him to the pound. But the people don't pay any attention to her. They're too busy trying to lure Rufus out from where he's cowering under the seat. Ruby doesn't blame him—these people are ridiculous. And so she makes a snap decision that leads to a surprising choice. She decides to protect Rufus. She tells them the dog isn't who they think he is, she's sorry to disappoint them—it happens all the time since that other lookalike dog got so famous. So instead of letting Rufus go, she takes him. To save him from the crazy dog people. And also, on a deeper level, if she didn't have to deal with the dog, she'd be more lost than she already is. Rufus offers her an odd kind of comfort in her moment of need—although it's not normal dog comfort (of course). And now Ruby has to find a place to write, and to hide with the dog . . .

Bingo! Ruby is now on the road; she can't go home; she has to deal with the dog, so she can't just hibernate somewhere; and seeing Henry in the hospital, hearing Frances talk about how she ruined his life, made her vow to write the script if it's the last thing she ever does. Oops, that's me. I got carried away filling in the blanks of the story—which I can do because the foundation is there and the layers are building and it's becoming easier and easier to see and to *feel*. Which is, of course, the whole point, and the fruits of Jennie's labor.

SCENE #4

ALPHA POINT: Ruby goes to the hospital and learns Henry won't recover

FRANCES SUBPLOT: Henry's mother asks Ruby to leave and not return

	CAUSE	EFFECT
	What happens	**The consequense**
THE PLOT	• Ruby drives to the hospital on a wave of magical thinking: Henry will have awakened from his coma, and everything will be okay. • Ruby drives like a maniac; Rufus, scared, crawls under the passenger seat. • Ruby, blind with panic and grief, forgets Rufus entirely. • Ruby rushes into Henry's room to find his mother, Frances, holding his hand, crying.	• Ruby learns that Henry will not wake up or recover. • Henry's mother, Frances, is deciding whether to keep him on life support or let him go. Ruby has no say in the matter. • Frances blames Ruby for everything—Henry wasted his life waiting for her to come around. She asks Ruby to have the common decency to leave a mother to her grief.
	Why it matters	**The realization**
THE THIRD RAIL	• Ruby's true feelings for Henry—he's the only person she feels deeply connected to—surface, blinding her to what she's always told herself about their relationship. This sets her up for the real grief to come. • Ruby is trying to outrun the pain—it blinds her to everything else in the moment—the show, Nora, Clementine, and Rufus.	• In the face of Frances's raw pain, Ruby realizes she is more alone than she ever thought she could be. • Ruby wonders if Frances is right and she did ruin Henry's life. • Ruby is on the verge of giving up, letting Nora take over, and letting Clementine write the finale. **And so?** With nowhere else to go, Ruby decides to go home.

And now with the fifth Scene Card on the following page, Ruby, and Jennie, are now on their way! Here's the surprise: Jennie has not only been nailing down every layer of her novel, she's also written more of it than she thinks. I'm not talking about just her opening and "aha!" moment scenes. A lot of the exploration she did, and the backstory she wrote, will be part of her novel, either as full-on flashbacks, or as the thoughts that will go through Ruby's head as she navigates the quickly escalating plot that Jennie has begun to craft. This is exactly how you bring a novel to life. It's the work you've already done that guarantees that your novel will be rich and evocative and logical and whole.

WHAT TO DO

While the Scene Cards you have for events further down your novel's time-line may still have nothing more than their Alpha Point noted, chances are Scene Cards #2 through #5 are probably just about complete. If not, do what Jennie just did, asking questions and filling in any logic gaps that leap out at you until these cards are ready to transition into an actual scene.

What Now?

It's finally time to write Scenes 2 through 5 and do a bit of rewriting in your opening scene. Because I'm willing to bet you've discovered a lot of new, juicy info that you'll want to strategically weave into the opening so it can play forward. And that's as it should be. You will continually be hunting for new information and incorporating it into your novel. Every time you hit a place where you're not sure what happens next, or why something happens, stop. Review what you know, and flip through your cards—the ones already in your timeline, and those that haven't yet found their place. If you do not find the answer there (and chances are you won't, especially

SCENE #5

ALPHA POINT: Ruby protects Rufus from his fans

	CAUSE	EFFECT
	What happens	**The consequence**
THE PLOT	• When Ruby comes out of the hospital, several people surround her car, trying to lure Rufus out from under the seat. • Ruby realizes she could let Rufus go, say she found him running loose and was heading to the pound. • Ruby notices the people are frightening Rufus, who's cowering under the seat, and looking at her for safety.	• No one has gotten a good look at Rufus yet—but they're so over the top trying to get his attention that Ruby decides to protect him. • Ruby tells them he's a lookalike, sorry—her life's been made miserable ever since that damn Internet dog started, what's the word? Trending. • The people back off; Rufus looks at Ruby, grateful, and thankfully doesn't bark. (Perhaps someone recognizes Ruby?).
	Why it matters	**The realization**
THE THIRD RAIL	• Ruby feels abandoned, friendless, alone, and when she sees the world coming in on Rufus, she suddenly feels protective of him. • It feels good to be angry instead of weak; to take action instead of flee. • Rufus keeps her in the game, gives her something to do.	• Ruby realizes that no matter how their show ends, he'd want her to write it herself. • Ruby realizes that she can't go home, because the confrontation with Nora would keep her from writing. • Ruby realizes she's stuck with Rufus for a while longer, and so she needs to keep a keen eye out for cops, which is surprisingly exhilarating. **And so?** Ruby goes on the lam with Rufus.

early on), it's time to dive into the past again for more inside intel. The good news is that the more you know about your story and your protagonist, the keener your sense of what you're looking for. Because—yep—specifics keep on begetting specifics.

By the time you've written those first five scenes, and are ready to start on the sixth, you will have dozens of Scene Cards in your Development Folder. Keep them in a rough chronological order: this is your fledgling cause-and-effect trajectory. Remember, if you're not sure where a particular card will go—for instance, your protagonist told a lie before the novel begins that you know will be revealed before it ends, but you're not sure exactly when—put that card at the front of your trajectory. That way it won't get lost or forgotten, and every time you go through your cards you'll be reminded not only that this scene is still to come, but of the layer of reality it represents for your protagonist. Because when it comes to secrets and lies, the more damning the lie, the more we think about it, and the more it drives our behavior. Point being: It's a ball in play, and you don't want to lose sight of it.

Stories Grow in Spirals

Make no mistake, all this back-and-forthing will continue throughout the entire process—up to the moment you decide your novel is ready to go out into the world. Expect to be circling back to the beginning of your novel to strategically, often surgically, layer in new info, new setups, and new storylines. Best-selling author Harlan Coben talks about this process, saying that every seventy-five pages, he goes back to the beginning of the novel he's writing and reads forward, editing, adding, subtracting—making sure every layer is firmly rooted in place, beginning on the very first page.

Why? The butterfly effect. Sure, popular culture has exaggerated the notion of the butterfly effect a wee bit (I mean, really, go back in time to the Stone Age, add one measly butterfly, and today we're all wearing red socks, there's no Internet, and unicorns exist?), but in a novel the butterfly effect—the notion that one itty-bitty change in the past can have a massive effect as it plays forward—couldn't be more relevant. As you write forward, if you change something, or add something new—and you will—that change will not only ripple into the future, but it will affect the past as well.

Remember, nothing is written in stone; it's sculpted in clay. Expect things to change, because with each layer you add, the ones beneath will shift a little to accommodate it. The good news is that because you've tracked each subplot throughout your blueprint, you'll know precisely what, and where, those changes are, so you can be sure your story logic still holds, and rethink it where it doesn't. The point is, change is constant. Don't let it throw you. This is part of the process. The Scene Cards will help you keep track of all this. But here's the real, amazing, time-saving, you-can-breathe-a-great-big-sigh-of-relief truth: you are well on your way to the first draft of an actual story, not a 327-page storyless novel about a bunch of things that happen. Will there be some rewriting? Of course, there always is. But it will be surgical—that is, you'll go back and layer new storylines and new info into what you already have, rather than having to rewrite those 327 pages starting on page one. That alone is worth the price of admission.

Three Secrets for Making the Invisible, Visible

This is the stage where it begins to feel as if your characters have taken on a life of their own, and they're telling you what might happen next, and they're not shy about how they feel about it. Don't be fooled: it's not *them* talking, it's coming from deep inside you—and the life that *you* gave them. Savor that! But don't fall into the one last trap that lures so many writers to their downfall: imagining that because *you* can read your protagonist's mind, your reader can too.

Your job from here on out is to make the invisible, visible. That is, how your protagonist sees the world, how she feels, how she makes sense of what's happening in the moment. This is no small feat. There are three secrets that will help you do just that—secrets you can *only* employ because of the work you just did. Use them, and your novel will hook, hold, and enthrall your reader from start to finish.

Your Protagonist Must Draw a Strategic Conclusion from Everything He or She Notices

When it comes to your protagonist, you can, in fact, read his mind. The trick is to give the reader the same experience. You'd be surprised how many writers don't, not because they're withholding on purpose, but because they don't know that the info *isn't* there. It's a strange optical illusion: because you already know so much about your protagonist, you have the uncanny ability to look through his eyes as he navigates the scene you're writing, making it astonishingly easy to forget that your reader can't do the same. So as you watch the events unfold you are, in fact, reacting to it the same way your protagonist would—you're thinking what he'd think, his memories are rushing through your mind, and it's thrilling, because as a result you're feeling exactly what he would feel. You are one! Ironically, because all of that is *already* right there in your head, you forget to put any of it into his. Luckily, the solution is surprisingly simple. You have to put it onto the page.

To wit: Your protagonist needs to react internally to everything that happens, in the moment, as he struggles to make sense of it and so bend it to his advantage. This means he thinks about what is happening. Not in the abstract. Not in long, rambling stream-of-consciousness musings. But with urgency, in service of a decision that must be made right now. The rule is this: he can't notice or comment on anything—even if it's just a description of what someone is wearing—unless he then draws a strategic conclusion *that affects what he's doing or how he interprets what's happening.* And he must do this every minute of every day—just like we do.

Think about your own life. You're constantly scanning your external surroundings for strategic meaning. *Is this safe? What should I do? What did he mean by that?* Where might danger (or delight) lurk? Whether you're standing on the platform at 31st Street in Astoria waiting for the Q train, in the middle of a battle in Gettysburg, or floating in space trying to get Hal to open the damn pod bay doors, you're constantly on the lookout for useful intel, the better to safely navigate that long walk home. Why would your protagonist be any different?

You Must Get Emotion onto Every Page

Let's make one thing very clear: although the reader needs to know how your protagonist feels at every turn, that does *not* mean you need to *tell* us. As in, *Thinking of life without Henry made Ruby feel gut-wrenchingly bereft*. When conveying how a character feels, the answer isn't happy, sad, angry, jealous, or bereft, gut wrenching or otherwise. Why? Because those things are, you guessed it, general. They are the "What," when, as always, what the reader wants to know is, *Why?*

The secret is this: the emotion emanates from how the character makes sense of what's happening, rather than mentioning the nearest big emotion that sums it up. Your goal isn't to tell us how they feel, so we know it intellectually; it's to put us in their skin as they struggle, which then *evokes* the same emotion in us. That feeling will be subtle, nuanced, and layered. And, most likely, ineffable, which is why you want to shun those massive generalizations. A mistake we make both in literature and in life is the tacit assumption that all emotions are defined by—and limited to—the catchwords we've come up with for them. Emotions are far more fluid and calibrated, and can't really be extrapolated in words (or, perhaps, it's just that we don't yet have the language for it).

Want an example? How about the following passage from Celeste Ng's debut novel, *Everything I Never Told You*. The scene takes place in 1966. Marilyn, a white housewife, has just learned of her estranged mother's death. They hadn't seen each other since Marilyn married her Asian college professor, James Lee.

> By then she had not spoken to her mother in almost eight years, since her wedding day. In all that time, her mother had not written once. When Nath had been born, and then Lydia, Marilyn had not informed her mother, had

not even sent a photograph. What was there to say? She and James had never discussed what her mother had said about their marriage that last day: *it's not right.* She had not ever wanted to think of it again. So when James came home that night, she said simply, "My mother died." Then she turned back to the stove and added, "And the lawn needs mowing," and he understood: they would not talk about it. At dinner, when she told the children that their grandmother had died, Lydia cocked her head and asked, "Are you sad?"

Marilyn glanced at her husband. "Yes," she said. "Yes, I am."[1]

There was nothing in the passage that mentioned how Marilyn actually felt, and yet everything in the passage conveyed it. It ends with a send-up of the word "sad." Sad is the dictionary definition of what we're taught to expect in situations like this. Sad is the one thing Marilyn doesn't feel; it's also the word she hides behind, to protect both Lydia and herself from the far more complex, and potent, emotions she is actually experiencing.

You Must Stay in Your Protagonist's Subjective Mind-Set

What stokes a story's momentum isn't simply what happens; it's what it costs the protagonist *internally* to make the decisions that drive the external action. Decisions that, on the surface, might *appear* objectively irrational, but are completely logical to your protagonist, because they're based on his subjective beliefs.

Here's an interesting thing to mull over: When we look at someone else—like, say, the hoarder in the apartment upstairs who hasn't thrown out a single magazine or newspaper since 1988 (and she subscribes to them all, having never gotten the hang of this newfangled Interweb thing)—we think, *I can't believe she does that, how irrational.* But, and this is the point, it's irrational only from the outside looking in. If you asked *her* about it, she'd no doubt have a perfectly logical reason for it, one that makes so much sense to her that she can't understand why you'd even question it.

To your protagonist, everything she does makes complete sense, given the internal set of rules she lives by. She's hardwired that way. We all are.

That's why it's not enough for you to know how she sees the world; you have to make sure that it's there on every single page.

Because here's the thing: that subjective lens through which your protagonist sees the world is a constant. It sounds obvious, but what does that mean, exactly? I ask because it's something writers often lose sight of. Their protagonist will be faced with something utterly life-altering, and then, in the midst of it all, spend hours doing errands, gabbing with friends, and exercising as if it's just a plain old day that could take place any time, anywhere, to anyone. So if you plucked that scene out of the novel and read it by itself, you'd have no clue that the protagonist was struggling with anything at all. Largely because she isn't.

It's like if Jennie wrote a scene in which Ruby decided to take a leisurely stroll along the crowded beach with Rufus, playing fetch and chatting up strangers, without once worrying that someone might recognize him. That would make you think that the whole Rufus-is-totally-recognizable thing wasn't nearly as serious as it seemed. Point being, we take our cue from the protagonist. If she's not worried about it, then why should we worry? After all, we know from personal experience that when something genuinely horrid is going on, it's always with us no matter how much we pretend it isn't. It not only filters everything we see, it tells us what to look for.

That does not mean that characters might not do things that are totally out of character, or even illogical; it just means that when they do it, you need to have established a credible "Why" beforehand. For instance, we know that in Scene #4, Ruby is going to drive in a panic to the hospital to see Henry, and then leave Rufus in the car, in plain view of anyone who walks by. Sounds crazy. Until you realize that at that moment Ruby *is* crazy—with grief and fear. She has her eyes on only one thing: getting to safety, which means Henry. She is blind to everything else.

Let's take a last dip into brain science to see why. You may have heard of psychology professors Christopher Chabris and Daniel Simons's gorilla experiment. It's an eye opener about an eye closer. They asked a group of students to watch a video of two teams passing around a basketball. One team was dressed in white, the other in black. Students were asked to count the number of times the ball was passed by the players wearing white, and to ignore the passes by the team in black. They did. But 50 percent of them failed to notice that halfway through the video, someone wearing a full-on gorilla costume walks through the shot, stops, beats her chest, and then wanders off. The gorilla is on screen for nine full seconds. This experiment has been repeated countless times, and the results are always the same. What astounded the researchers even more than the fact that half of the people consistently miss the gorilla, is that when told about it, they don't believe it even happened. *I couldn't have missed it!* is a common refrain—even, *You must have switched the tape, no way the gorilla was there the first time you showed it to me!* But it was. This is known as inattentional blindness.[2] The point is, when your protagonist is focused on the problem that drives her, it remains front and center, editing out what is irrelevant, so she's quite literally blind to it, while selecting and coloring everything she *does* see.

This neatly exposes the real meaning of a quip that, up to now, seemed like plain old common sense: "Seeing is believing." Who knew it was largely meant ironically? Because as far as those people who didn't *notice* the gorilla were concerned, it simply wasn't *there*. After all, they knew what they saw, thus inadvertently proving that while seeing might be believing, believing doesn't make it true. But here's the rub: seeing makes it *feel* true, and that's the point. We see the world through what feels true to us, and that dictates both what we notice and the meaning we read into it.

Want to feel it yourself? Imagine that you wake up and realize you have a hangnail on your left pinkie. Panic seizes you, because you instantly know what that means: you have the dreaded hangnail virus, and it's one of the deadly ones; you're sure to be a goner by midnight.

Even if you then went about your day exactly as you would have sans dreaded disease, nothing would be the same. You'd notice different things and you'd read different meanings into what used to be familiar—going to the market, walking the dog, calling your kids. Everything would be viewed through the lens of *this is the last time*, and that would profoundly change it. Of course, this whole scenario is ridiculous. There's no such thing as hangnail virus (you can stop your Google search now). Your fear would have been totally, completely irrational by real-world standards. Does that change anything? Not a chance. Because we see the world through our own subjective lens come hell, high water, and a thousand rational explanations of why we're wrong.

The same is true of your protagonist. Whatever is driving her, whatever she's worried about, it's always there, urging her forward, directing her attention, defining what she notices and what she doesn't, what she remembers, and the conclusions she draws—and as you know, she's *always* drawing conclusions. When it comes to Ruby, it means that her grief will be with her all the time. Thoughts of Henry will haunt her, soothe her, sting her, and drive her. They will color how she sees—and interprets—everything that happens from the first page to the last.

And that's not all. From the moment each layer is added, it too will be constantly present, as the stakes mount and she struggles with what to do. For instance:

- Once Nora threatens to sell her house, Ruby's fear that she'll lose her home will be ever present.
- Once she learns the deadline for the script rewrite is immutable, panic about what will happen if she doesn't deliver will be ever present.
- Once she snatches Rufus, fear that she'll be caught with the dog will be ever present.

As Jennie weaves these elements in, they'll meld together, becoming part of the omnipresent lens through which Ruby views the world, informing every scene she's in. Notice too that each one also has an external counterpart, an unavoidable deadline with a countdown clock that's rapidly ticking toward a consequence that will have a profound effect on Ruby's life. In other words, there's no escape. And so there won't be a moment when Ruby isn't keenly aware of what's coming her way, *even* in those moments when she finds respite and solace.

In the beginning, adding in these elements may feel heavy-handed—like you're working with a blunt tool. That's fine. The point is that a scene is not just something that happens. (Sound familiar?) A scene is designed to move your plot, along with myriad subplots, forward, powered by your novel's third rail—and yes, I am referring to *all* of those things, all at the same time. Remember, a single scene doesn't take one step forward; it's akin to taking eight different steps forward, all in unison. That's why at first it will feel very rough, but as you add in layers, they'll begin to meld as the scene unifies and gathers nuance. The result isn't mathematical. It's exponential. You go from one layer, to two, to three, and then, suddenly, your novel has taken on the depth and breadth of life itself.

Here is an example of Jennie doing this work. She wanted to work on the second scene, when it dawns on Ruby that "borrowing" a dog is the perfect solution to her problem. Jennie attempted to weave in all the constantly present realities *and* show how they move the story forward. It's raw. This scene might go through several iterations before it's all the way there.

The scene opens with Ruby having decided that the way to convince Nora to leave her alone is to pretend she's eating well. Yeah, *that* would work.

As soon as the sun came up, I showered and put on real clothes for the first time since the accident. I was waiting at Whole Foods when the store manager came to the door with her key to let in the early morning supplicants. I grabbed a cart and began to load it with organic spinach, rosemary sourdough bread, fresh garlic hummus, and seasonally appropriate fruit. I went for the most expensive option of everything offered, and finally steered into the meat aisle, thinking I would get stuffed chicken breasts or a marinated roast—something with gravitas I could stick in the oven so that when Nora arrived the house would smell of someone taking care of themselves. I pulled my cart behind a woman with a small white dog tucked into a bag that was slung on her shoulder.

"A pound of the organic grass-fed hamburger," she said. The dog poked its head out and she began to murmur.

"That's right, Bruiser, Mama's getting your favorite."

I knew how much some dogs were pampered, but I had never imagined anyone buying grass-fed beef for their dog.

"It eats organic meat?" I asked.

"She absolutely refuses kibble," the woman said, "and you can't really blame her, can you? It's packed with preservatives and added fillers, and the meat they use to make that stuff—arghk." She made a sound of disgust and scratched the dog behind its ears. "It's the scraps from the slaughterhouse floors. It's enough to turn your dog vegetarian."

"Dogs can be vegetarian?" I asked—not because I doubted the answer, but because I wanted to hear this woman explain how the descendants of wolves could survive on tofu. It seemed suddenly important.

She laughed. "Of course! You just have to give them enzymes so they don't have gas from all the fiber." She took the package of meat the butcher handed her, cooed again to her dog, waved, and headed off toward the eggs. I supposed the dog liked poached eggs for breakfast.

"Can I help you, ma'am?" the butcher asked.

It took me a second to realize he was speaking to me. Ever since Henry's accident, the world seemed to move at the speed of light, and I was constantly scrambling to make sense of it.

"Yes," I said, and forced a smile to buy a little time.

I glanced down the aisle after the woman with the dog. Suddenly something began to take shape in my mind. Sharon's receptionist thought I should get a dog, and the woman who cut my hair, and the mailman who brought nothing but junk and bills. Nora had been suggesting I get a dog all my life—she, the veterinarian, for whom doglessness was evidence of everything that was wrong with my life. I had no husband, no family, no dog, and my entire ability to connect with the human race was therefore suspect. "You should get a dog!" It was the knee-jerk reaction people gave to single women and middle-aged women and depressed women and women in pain, and I was now, all of a sudden, all of those things.

Forget the organic salad and roasting meat. If Nora thought I had a dog, she would stop worrying about me completely. If she thought I had a dog, she would leave me alone, and then the only thing standing in the way of writing would be figuring out how to do it without Henry pacing around the table, gazing out the window, and chewing on a pen cap. I didn't have to buy a dog, or adopt one. I just needed a dog for a few hours.

I turned to the butcher. "A pound of the grass-fed hamburger, please."

Why This Scene Works:

- Did you notice that what Ruby is always aware of is the unavoidable change she's facing—Henry's absence, Nora's threat to sell the house, the script deadline?

- Did you notice that Jennie never told us how Ruby felt and yet we were viscerally aware of it?

- Did you notice how much of this scene is backstory? Ruby's memories are the yardstick she uses to justify the rather dodgy plan she's just come up with.

- Did you notice that every "sensory" detail Jennie gave us was in service of the story itself? She didn't tell us about the kind of food Ruby was buying simply so we'd know what was on her shopping list; she told us *why* she was picking such specialty items: to impress Nora.

- Did you notice Ruby's thinly veiled disdain for how people go overboard pampering their pets? She didn't have to say it outright; it was implied in the questions she asked the woman at the meat counter.

- Did you notice that Jennie gave us Ruby's train of thought, culminating in her decision to nab a dog? Thus we don't just know *that* she's going to grab a pooch, we know *why*.

- Did you notice that Jennie clued us in to Ruby's agenda, so we know what Ruby expects will happen as a result of nabbing the dog? To wit: Nora will leave her alone. Thus we can anticipate what might happen next, knowing one thing for sure: no good will come of Ruby's plan.

The reason Jennie could accomplish so much in a single scene is because she was able to slip into Ruby's point of view and see the world as Ruby saw it. And, just as important, she then put it on the page so we could see it too.

WHAT TO DO

As you approach each scene, gather everything you know about your protagonist and their subjective worldview at the moment. What are they most worried about? How will that affect their judgment in this scene? Be sure you've thought about each layer, each ticking clock.

All those strategic conclusions we were talking about in Secret #1, the conclusions that your protagonist needs to be continually drawing? Understanding what she's struggling with will tell you what those conclusions will be, and what she'll do as a result. That's what unites the story and the plot, and what moves them ever forward.

And Finally . . .

Not to seem like a mother hen, but I know that when you're out there writing, no matter how focused you are on getting the story in your head onto the page, there will be times when the force of an idea sweeps you away. And everything you've learned will vanish in the haze.

When that happens, and it will, I want to remind you of the two simplest, yet most essential writers' tools you'll ever possess:

- Ask "Why?" of everything, and don't stop asking until you've chased it down to its most story-specific, flesh-and-blood, "close your eyes and you can see it unfold" origin and there is not another "Why" to ask.
- Ask "And so?" of everything. And so, why does my reader need to know this? And so, how does this move the story forward? And so, what will happen as a result? In other words: What. Is. The. Point? Harsh as it sounds, this question is your most stalwart friend; it will never pull punches, and it will often reward you with a revelatory, story-specific insight that you'd never have gotten to otherwise.

You now have all the tools you need to write a riveting novel capable of triggering a dopamine rush in your reader's brain that will make her forget everything else: the series she was about to binge watch, the big meeting tomorrow at work, and even better, all those nattering doubts about, well, everything, that seem to descend like a swarm of mosquitoes the later it gets. Instead, she'll get a mini-vacation, a moment of solace, of sanctuary as she takes refuge in your novel, all the while picking up inside intel into what *really* makes people tick. And when she finally turns that last page, she will emerge changed, because whether she knows it or not, her view of the world will have shifted. It might shift in a small way, or it might be a really dramatic change that impacts everyone and everything around her. After all, riveting novels have been known to change the world itself.

For instance, do you know what is often cited as a major reason for the success of the civil rights movement in the 1960s?

To Kill a Mockingbird.

In fact, a 1991 survey by the Library of Congress Center for the Book found that *To Kill a Mockingbird* was rated second, behind only the Bible, in books most often cited as making a difference.

Oprah Winfrey calls it "Our national novel."[3] Former First Lady Laura Bush said, "It changed how people think."[4]

How? By changing how they *felt*.

Because the only way to change how someone thinks about something, is to first change how they feel about it.

You have the power. Now go use it.

ACKNOWLEDGMENTS

Everything is easier said than done. Except getting into trouble, which, maddeningly, is actually *easier*. That's why I owe an immense debt of gratitude to the many people who kept me out of trouble in the writing of this book, and made it much better than it ever would have been without them.

I'm forever grateful to the friends, family, and fellow writers who in one way or another helped me form, shape, try out, and hone every idea presented here: Lynda Weinman, Michelle Fiordaliso, Annie Jacobsen, Bryce Dallas Howard, Kelly Carlin, Jane Praeger, Sara Cron, David Benton, Nancy Blaha, Cynthia Anderson, Christine Fletcher, Rebecca Pekron, Jennifer Tracy, PJ Arnn, Anne Childers, Kelly Clark, Camille Kress, James Ballard, Jennifer Appenrodt, Jessica Yellin, Margaret Tsubakiyama, and especially Amit Chatwani, whose refusal to compromise on any aspect of his novel, no matter how much work it entailed, forced me to dig deeper than ever.

Thanks to Frank and Medina Fredericks, in whose cozy apartment in Astoria I wrote the bulk of this book in the dead of a very snowy winter.

Endless thanks to the people who read the manuscript in various stages and generously took the time to give me much needed feedback—Colin Kindley, Joni Templeton, Mark Rovner, Caroline Leavitt, Laura Franzini, and Carlyn Robertson, whose early spot-on advice was invaluable. Special thanks to Chris Nelson, for his unerring eye and unwavering support.

This book is far stronger thanks to John Z. Nittolo, the maverick superintendent of a small school district in Newton, NJ, who brought me in to help them incorporate story into how they teach writing from kindergarten on, and Tara Rossi, the inspiring young literacy coach I worked with at Green Hills School. Thanks also to the kids themselves, who astounded me with how quickly, and eagerly, they embraced story, and how much they had to say.

A big thanks to the smart, engaged, supportive community at Writer Unboxed. You guys keep me on my toes. Deepest thanks to editor-in-chief Therese Walsh, whose grace, generosity, and strategic thinking have helped create one of the most inspiring writers' hangouts on the Web. We'd be lost without you, especially me.

Jennie Nash, what can I say? This book wouldn't exist without your indefatigable, insightful, beautifully brutally honest coaching. Nor would it exist without your having so generously been willing to candidly expose those first thrilling, agonizing, frustrating moments of wrestling an idea onto the page. I love your novel.

Endless undying gratitude to my daughter, Annie, who willingly read anything and everything without ever letting on that the hundredth read might be a tad excessive. I'm forever humbled by my son, Peter, whose love of story is unparalleled, and whose keen insights continually deepen my own. Never stop pushing me.

Once again, profound thanks to my brilliant agent, Laurie Abkemeier of DeFiore & Company, who has the uncanny ability to make everything work out even better than expected. I remain in awe. Huge thanks to my whip smart editor, Lisa Westmoreland, my copy editor Kristi Hein, and the savvy team at Ten Speed Press. You guys rock!

Everlasting thanks to my lifelong best friend, Don Halpern, who's there in just about every memory I have, and finally, deepest heartfelt gratitude to my husband, Stuart Demar; you make me feel safe enough to put on the blinders and write. Might not sound like a lot, but it's everything.

ENDNOTES

Introduction

1. F. O'Connor, *Mystery and Manners: Occasional Prose* (New York: Farrar, Straus and Giroux, 1970), 66.

Chapter 1

1. D. Gilbert, *Stumbling on Happiness* (New York: Vintage, 2007), 76.

2. J. Gottschall, "The Storytelling Animal," filmed March 2014, TEDxFurmanU video, 12.26, posted May 2014.

Chapter 2

1. E. L. Doctorow, "The Art of Fiction No. 94," interviewed by George Plimpton, *The Paris Review*, Winter 1986, Issue 101, accessed online, http://www.theparisreview.org/interviews/2718/the-art-of-fiction-no-94-e-l-doctorow.

2. A. Lamott, *Bird by Bird* (New York: Anchor, 1995), 20.

3. Horace, *Ars Poetica*, l. 147.

4. D. Aligheri, translated by J. Ciardi, *The Divine Comedy* (New York: NAL, 2003), ix.

Chapter 3

1. Museum Kids, *The Mixed-Up Files Issue* (New York: Metropolitan Museum of Art, 2001).

2. S. King, "Best Selling Author Stephen King Talks About *Under*

the Dome," Simon and Schuster Books video, 2:11, October 2009, https://www.youtube.com/watch?v=GEUj_klOhd4.

3. J. Didion, "Why I Write," *New York Times Book Review,* December 5, 1976.

4. S. Johnson, *The Beauties of Samuel Johnson, Ninth Ed.* (London: G. Kearsley, 1897), 2.

Chapter 5

1. W. Berger, *A More Beautiful Question: The Power of Inquiry to Spark Breakthrough Ideas* (New York: Bloomsbury USA, 2014), 40.

2. "Louie Works Through a Chain of Whys," https://www.youtube .com/watch?v=8idwyuVJ4ug, posted August 2010.

3. I. Glass, *This American Life*, "552: Need to Know Basis," originally aired March 27, 2015. Transcript: http://www.thisamericanlife .org/radio-archives/episode/552/transcript.

Chapter 6

1. B. K. Bergen, *Louder Than Words: The New Science of How the Mind Makes Meaning* (New York: Basic Books, 2012), 19.

Chapter 7

1. K. Oatley, "The Science of Fiction," *New Scientist,* June 25, 2008, 42–43.

2. http://www.mtvu.com/shows/stand-in/trey-parker-matt-stone-surprise-nyu-class/, 4:30, posted October 4, 2011.

3. S. J. Watson, *Before I Go to Sleep* (New York: Harper, 2013), 156.

Chapter 8

1. J. F. Kennedy, *The Letters of John F. Kennedy,* ed. M. W. Sandler (New York: Bloomsbury Press, 2013), 5.

Chapter 10

1. J. C. Oates, interviewed by Brian Lehrer, *The Brian Lehrer Show,* WNYC, February 6, 2015, 18:16, http://www.wnyc.org/story/

joyce-carol-oates-take-tawana-brawley-case.

2. J. Irving, "Getting Started," in *Writers on Writing*, ed. R. Pack and J. Parini (Hanover, NH: University Press of New England, 1991), 101.

3. J. Irving, *A Prayer for Owen Meany* (New York: William Morrow Paperbacks, 2013), 3.

4. A. S. A. Harrison, *The Silent Wife* (New York: Penguin Books, 2013), 3–4.

5. D. Tartt, *The Secret History* (New York: Vintage Books, 1992), 7.

6. T. S. Eliot, *Four Quartets* (Boston: Mariner Books, 1968), 59.

Chapter 13

1. M. M. Waldrop, *Complexity: The Emerging Science at the Edge of Order and Chaos* (New York: Simon and Schuster, 1992), 47.

Chapter 15

1. C. Ng, *Everything I Never Told You* (New York: The Penguin Press, 2014), 80.

2. C. Chabris and D. Simons, *The Invisible Gorilla: How Our Intuitions Decieve Us* (New York: Harmony, reprint 2011), 6–7.

3. M. M. Murphy, *Scout, Atticus and Boo: A Celebration of To Kill a Mockingbird* (New York: Harper Perennial; reprint edition July 5, 2011), 3.

4. M. Puente, "'To Kill a Mockingbird': Endearing, Enduring at 50 Years," *USA Today,* August 8, 2010, http://usatoday30.usatoday.com/life/books/news/2010-07-08-mockingbird08_CV_N.htm.

ABOUT THE AUTHOR

Lisa Cron is the author of *Wired for Story: The Writer's Guide to Using Brain Science to Hook Readers from the Very First Sentence,* and the video tutorial *Writing Fundamentals: The Craft of Story,* which can be found at Lynda.com. Her TEDx talk, "Wired for Story," opened Furman University's 2014 TEDx conference, "Stories: The Common Thread of Our Humanity." She is a monthly contributor on the award-winning writers' website Writer Unboxed.

Lisa has worked in publishing at W. W. Norton, as an agent at the Angela Rinaldi Literary Agency, as a producer on shows for Showtime and Court TV, and as a story consultant for Warner Brothers and the William Morris Agency. Since 2006, she's been an instructor in the UCLA Extension Writers' Program, and she is on the faculty of the School of Visual Arts MFA Program in Visual Narrative in New York City. She is a frequent speaker at writers' conferences, schools, and universities.

In her work as a story coach, Lisa helps writers, nonprofits, educators, and journalists wrangle the story they're telling onto the page. She splits her time between Santa Monica, California, and New York, New York. She can be reached at wiredforstory.com.

INDEX